Cafe House 2

출판사 : 바른손
저　자 : 김 연 경
발행일 : 2024. 08. 20
주　소 : 경기도 파주시 월롱면 덕은리 216-17
가　격 : ₩98,000원

Cafe House 2

출판사 : 바른손
저　자 : 김 연 경
발행일 : 2024. 08. 20
주　소 : 경기도 파주시 월롱면 덕은리 216-17
가　격 : ₩98,000원

What is Bakery?

A bakery is an establishment that produces and sells flour-based food baked in an oven such as bread, cookies, cakes, donuts, pastries, and pies. Some retail bakeries are also categorized as cafés, serving coffee and tea to customers who wish to consume the baked goods on the premises. Confectionery items are also made in most bakeries throughout the world.

History

Usually small bakehouses were used in rural areas Baked goods have been around for thousands of years. The art of baking was developed early during the Roman Empire. It was a highly famous art as Roman citizens loved baked goods and demanded them frequently for important occasions such as feasts and weddings. Because of the fame of the art of baking, around 300 BC, baking was introduced as an occupation and respectable profession for Romans. Bakers began to prepare bread at home in an oven, using mills to grind grain into flour for their breads. The demand for baked goods persisted, and the first bakers' guild was established in 168 BC in Rome. The desire for baked goods promoted baking throughout Europe and expanded into eastern parts of Asia. Bakers started baking bread and other goods at home and selling them on the streets.

This trend became common, and soon, baked products were sold in streets of Rome, Germany, London, and more. A system of delivering baked goods to households arose as the demand increased significantly. This prompted bakers to establish places where people could purchase baked goods. The first open-air market for baked goods was established in Paris, and since then bakeries have become a common place to purchase delicious goods and to socialize. By the colonial era, bakeries were commonly viewed in this way.

Bakery & Dessert shop interior & Branding design?

When eating out, a good review isn't solely down to the food; a Bakery & Dessert shop interior plays an important role in the whole experience. Knowing how to create an environment that complements the menu and the space's interior architecture is no simple feat.

7 things to consider when designing a Bakery & Dessert shop.

1. Choose a Striking Colour Palette.

As with domestic projects, high up on the subject list for bar decor ideas is that of the colour scheme. "It's quite common for bar to be bold with their colour choices. In a sitting room or dining room there are just a few items of furniture and the colours chosen feel very much 'on show', in a Bakery & Dessert shopsetting, the volumes of furniture, accessories and light sources are far greater so there's a distraction from the colour. It's this that encourages the bravery and exploration in colour," Colour in a Bakery & Dessert shop can also become a huge talking point. In larger establishments, it acts as a tool to define distinct areas. The colour becomes an icon for the bar in question. Remember, also, with colour decisions to reflect on how they sit together in daylight and how they evolve when night falls and they rely on candle and lamp light for luminosity.

2. Master a Functional Layout.

"A Bakery & Dessert shop floor plan forms an immediate impression of ambiance. Keep tables spread out so that each feels secluded, and the guest experience will be an intimate one. "Tables packed tightly together on the other hand is a statement of conviviality and liveliness. Determining Bakery & Dessert shop layout requires a decision on what atmosphere is hoped to be established,' There's a purely functional aspect to the Bakery & Dessert shop room plan too though. The traffic should flow seamlessly. Bottlenecks are immediately apparent and cause an awkward distraction that guests will pick up on. A division between the Bakery & Dessert shop area is another question to be answered on the subject of layout. This will affect the feeling of formality or informality across the whole space.

3. Specify High-Grade Contract Furniture.

When a bar approaches its furniture choices as objects of function and nothing else, the entire experience becomes devalued. The importance of carefully selected bar chairs, tables and accent pieces is not to be underestimated. "It matters that not every element matches. That doesn't mean that every dining table and chair needs to be different, but that there are occasional pieces in the room to break up any consistency. A statement dresser or several elegant console tables adorned with decorative lamps or a vase are helpful here and serve as a reminder that Bakery & Dessert shop furniture extends beyond table and chair."

4. Select Show-stopping Lighting.

Arguably one of the most crucial aspects of Bakery & Dessert shop design, lighting ideas must be respectful of the fine line between necessary, task-style beams and ambient illumination. "In our homes, we consistently stress the need to layer the lighting throughout the space. It's no good having all of the lighting hung from above; it must drift down slowly from pendants to wall lights and lamps aplenty. A bar may be a commercial space, but guests want to be made to feel at ease in their surroundings as they do at home, so it's logical to follow the same lighting philosophy as you would in the home." Also acknowledges the effect that lighting has on each guest, considering what is the most flattering light at every angle. In the Harrods Dining Hall, they focus the lighting on the plate to showcase the food, with soft, low-level lighting for the diners to bask in.

5. Curate a Unique Decor Collection.

"If there's exquisite detail in the food, then there should be exquisite detail in the way the room is put together." Restaurant decor ideas, therefore, are a fundamental component in how the approach every commercial project. "The term 'finishing touches' is misleading. These aren't the bits to be simply added in at the end. Accessories and decorative touches are where you deepen the level of consideration in Bakery & Dessert shop interior design," Remember too that in the social media-centric world that we live in, carving out areas of appealing vignettes, complete with trinket boxes, decorative bowls and impressive floral arrangements, means your Bakery & Dessert shop design is all the more likely to become a must-visit and must-photograph location.

6. Display Personality-Full Artwork.

By extension of decor and accessories, displaying expertly curated arts and artefacts speaks volumes to guests. They serve as points of interest, they reinforce or subvert the overall interior design direction and they add warmth and texture to the space.
"Similar to how we light a Bakery & Dessert shop with as much thought as we do a home's dining area, bar wall design deserves as much thought as would be given in a residential project. Art adorning the walls and sculptures atop of tables are luxurious details that reveal the pedigree of the establishment. We always encourage this in our Bakery & Dessert shopinterior design projects, and it's an investment that our clients never regret."

7. Don't Neglect The Bakery & Dessert shop Bathroom Design.

It's not uncommon to hear creative types and interiors aficionados claim that to know the true dedication to design of a Bakery & Dessert shop you must check out the bathroom. This is where the same level of attention to detail seen in the main Bakery & Dessert shop either flails or flourishes. "The combination and contrast of materiality is so interesting in bathrooms. Marble and stone give a luxurious finish and contrast beautifully with timber, which can add warmth. Mirrored glass, brass and nickel all give reflection and a sense of luminosity." Lighting, too, is fundamental in this area of the bar, otherwise undoing all of the hard work in setting the perfect level of ambiance during dining.

Table of contents

008~021 Bakers Lounge Studio Ardete

022~037 Katsioulis Bakery SOLO Studio

038~051 Voila Pastry Taj Design

052~065 Dorbolò Gubana Visual Display

066~079 Panistas Bakery Zooco Estudio

080~089 Il Punto Gelato MODO architecture + design

090~105 Sacher Eck Wien BWM Architekten

106~115 Vyta Santa Margherita Collidaniel Architetto

116~123 Holiland bakery Universal Design Studio

124~133 Holiland Pink Store Das Design

134~141 Karnaros Bakery ARCHE Architecture & Design Lab Design

142~151 Dessert Moishi 4SPACE

152~161 IDA Bakery 4SPACE

162~179 Cloud & Co.-Doha Studio futura

180~197 Pasticceria Faiella Studio futura

198~207 Lisette Bakery Neowe design studio

208~219 Famiglia Ice Cream SET Ideas

220~231 Ofelè Pasticceria AFA Arredamenti, Arch. Simone Colombo

232~243 Picnic Salad & Bakery Kuudes

244~263 Lingenhel Store Destilat Design Studio

264~273 Las Ramblas Cursor Design Studio

274~287 Fuwa Fuwa Studio Yimu

288~295 Gelato Dal Cuore Hcreates design

296~303 Estrela Doce Bakery Sónia Triguinho

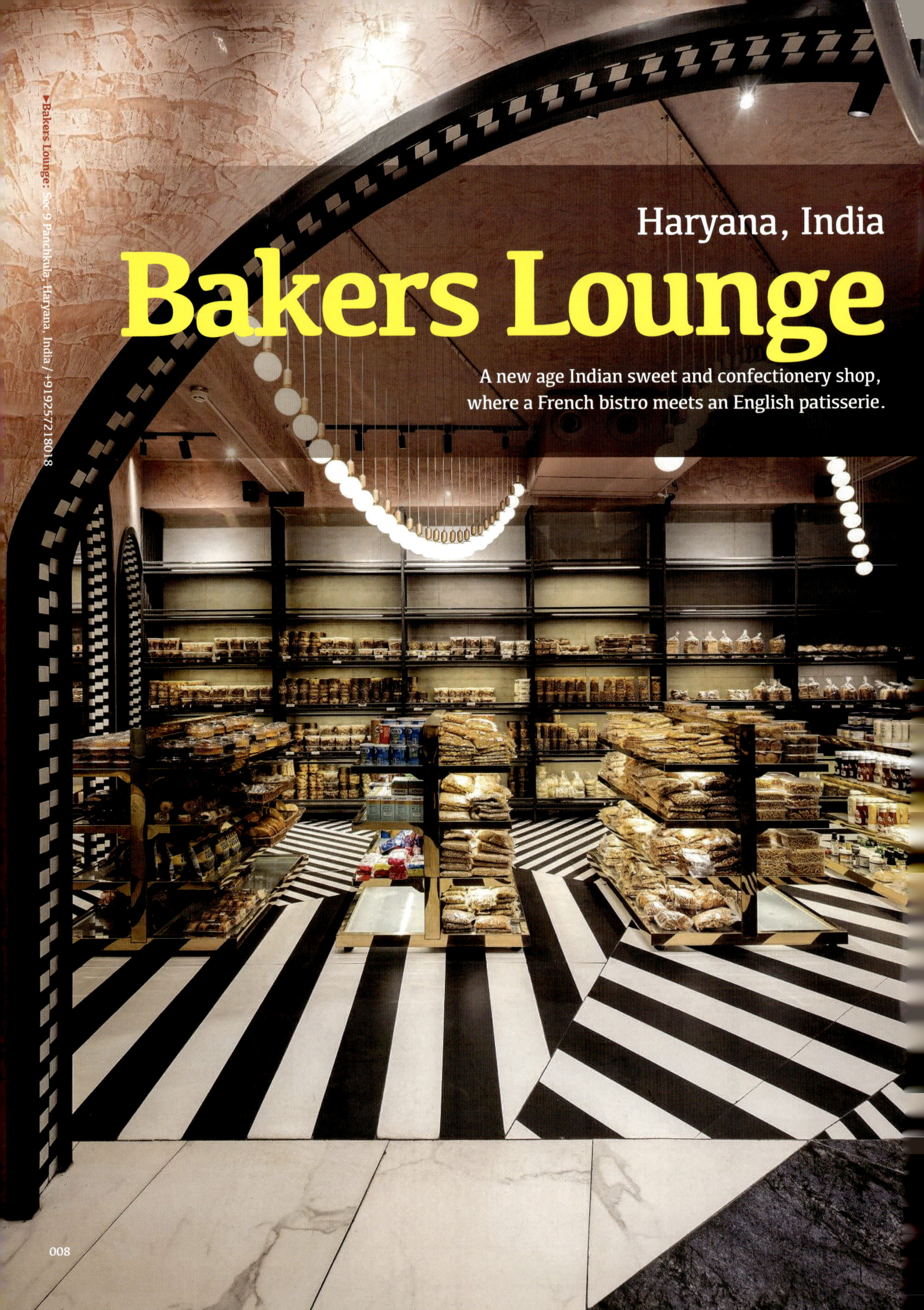

Bakers Lounge

Haryana, India

A new age Indian sweet and confectionery shop, where a French bistro meets an English patisserie.

In India, traditionally consumption and gifting of sweets or "mithais" are symbolic of celebration. It has a deep-rooted association with auspicious events and is an indispensable part of our socio-cultural heritage. Even today when you want to share a piece of good news with friends and family, it is expected that you would treat them to a box of delectable sweets. Over the past few years, the ambience of Indian sweet shops has become quite routine. Display shelves fashioned in similar patterns, unexceptional dining sit-outs and the usual flooring, nothing very exciting.

The design then became a voyage attempting to answer this question. How can a trip to the sweet & confectionery shop challenge assumptions of visual perceptions? Drawing inspiration from the versatile range of food products and contrasting flavours, concept evolution and development for Baker's Lounge became an enquiry: what would be the impression of a space that is somewhere between a French bistro and an English Patisserie? The objective then became a concoction for an unparalleled retail experience that would transcend all sensory barriers as visitors walked through the store.

In the current bustling scene of the food & beverage industry, emerges a new age sweet & confectionery shop that caters to a wide range of tastes; from a "Gulab Jamun" to a "Lemon tart"! Located in Panchkula, occupying 2500 square feet area on the ground floor of the tri-storeyed building is Baker's Lounge with a dining area that comfortably seats up to 25 people.

Project details

▶ **Design:** Studio Ardete
▶ **Homepage:** https://studioardete.com
▶ **Area:** 2265.3 ft²
▶ **Location:** Sec 9 Panchkula, Haryana, India
▶ **Photographs:** Purnesh Dev Nikhanj
▶ **Branding and graphic design:** Pravda design

About us

Believing that every problem has the capacity to inspire unique and creative solutions that motivate growth, the young duo of Badrinath Kaleru and Prerna Kaleru, founded Studio Ardete in the year 2010. Located in Chandigarh, India, the team at Studio Ardete aims to make each project they undertake, a study in logical and comprehensive designing. Their vision is to design spaces that would evolve into art, enriching the lives of people inhabiting them.

Not hesitating to take on even the most challenging of projects, they use modern construction and building methods to come up with designs that are at par with the most accomplished practices in the world. Ardete incorporates new techniques like parametric modeling and customized design elements, unique to each project they take on, striving to use materials available on site to make the projects sustainable. They work closely with the clients to understand their requirements and engage them in the design process, making them an intrinsic part of it.

The client envisioned creating minimalist yet impactful aesthetics. Thus it became essential to develop a visually de-cluttered retail experience for the customers while being considerate to the high footfall, especially during the festival season. Discernible zones had to be schematically planned catering to myriad offerings like packaged food products, baked items, 'to be ordered' food preparations and sweets. More importantly, it was essential to converge the zones to a common visual theme in an unassuming fashion.

An effort was made to recreate the bygone British colonial era vibe. A distinct visual tone was set with art deco reminiscent arches in chequered soffits, a kaleidoscope of monochromatic flooring and pastel blush ceilings. Suspended bulb light fixtures orchestrated a vintage symphony in the form of inverted arches while metallic gold accents seal the retro vibe of the store bringing in just the right amount of bling.

Baker's Lounge

sector 9, Panchkula

FRONT VIEW

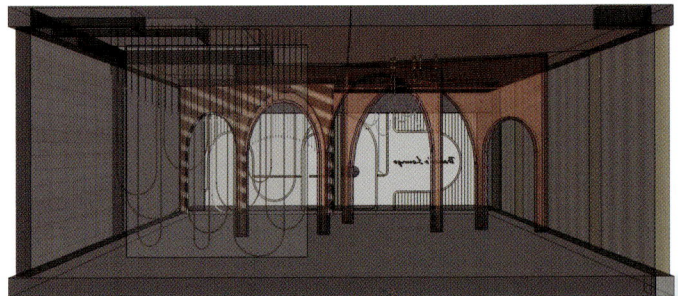
REAR VIEW

Different Radius Arches has been used in the form of **Beams**, **Chandeliers**, **Wall Claddings** and **Ceiling lights**.

The aesthetics progressed towards a symbiotic composition of layers of elements. Arches were specially crafted to depart from the dreary and usual, unidimensional display arrangement. They demarcated multiple axes that behaved as portals, highlighting multi-dimensional intuitiveness. Subconsciously distinct experiences were created at different hours of the day, evoking an element of surprise and wonder in the buyer. To accentuate the inflexion in the ceiling, soffits of the arches were clad in chequered tiles and highlighted with black profiles complementing the monochromatic flooring. Arches are employed as a recurring element in the visual scheme in the form of compositions of gilded membranes in the front display area, along the staircase and as a backdrop in the quick dining zone.

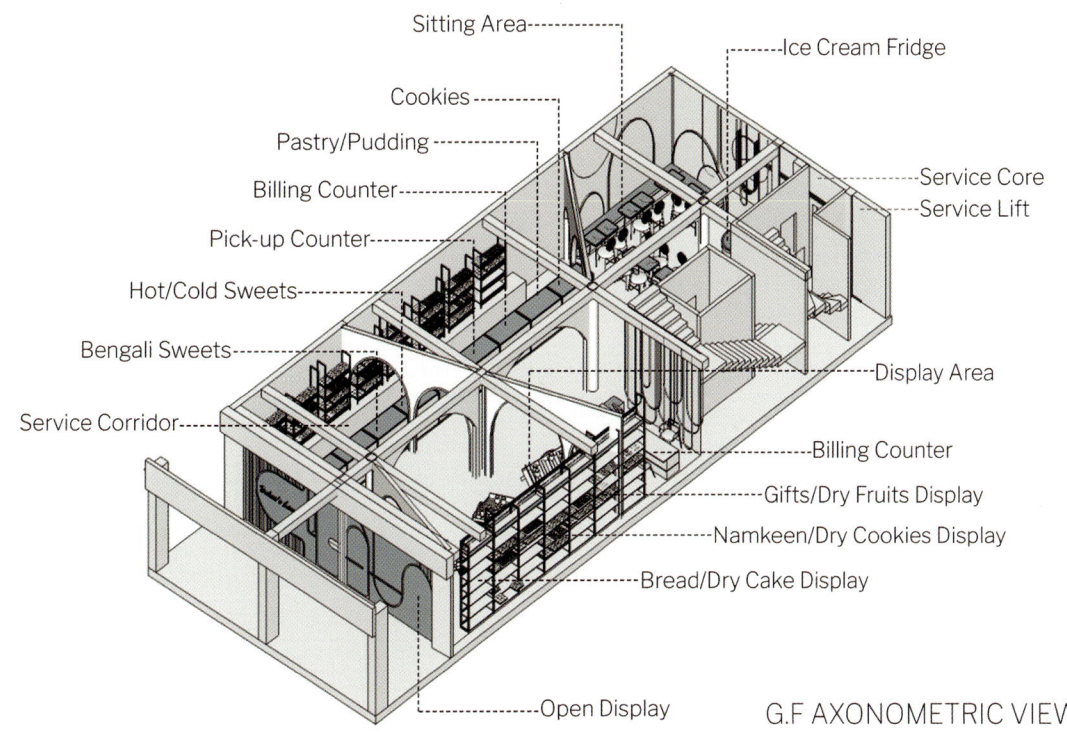

G.F AXONOMETRIC VIEW

- Sitting Area
- Cookies
- Pastry/Pudding
- Billing Counter
- Pick-up Counter
- Hot/Cold Sweets
- Bengali Sweets
- Service Corridor
- Open Display
- Ice Cream Fridge
- Service Core
- Service Lift
- Display Area
- Billing Counter
- Gifts/Dry Fruits Display
- Namkeen/Dry Cookies Display
- Bread/Dry Cake Display

Capturing the quintessential spirit of the 1920s, the pastel stucco marble textured ceiling is juxtaposed with monochromatic tiles. While the display racks along the light grey washed walls are kept muted, the free-standing shelves are finished in "aureate" edges for the classic "Midas touch". To facilitate the recall value of the visit, spherical and bulbous luminaries are suspended in a dramatic fashion, synchronising with the avant-garde spatial vocabulary dictated by the arches and their monochromatic character.

Embracing the vintage colonial times, fluted glass panels on the base of the display counters play a modest role while the combination of black, gold and white contributes significantly to enrich the visitor with a much deeper and larger experience. In the dining area, tufted tan leatherite seating, vault shaped backrests, inverted gilded membranes along with the monochromatic artwork on the walls weave a dynamic and vibrant ambience. Each element, distinct in nature and unique in function, unite to present themselves as a homogenous composition.

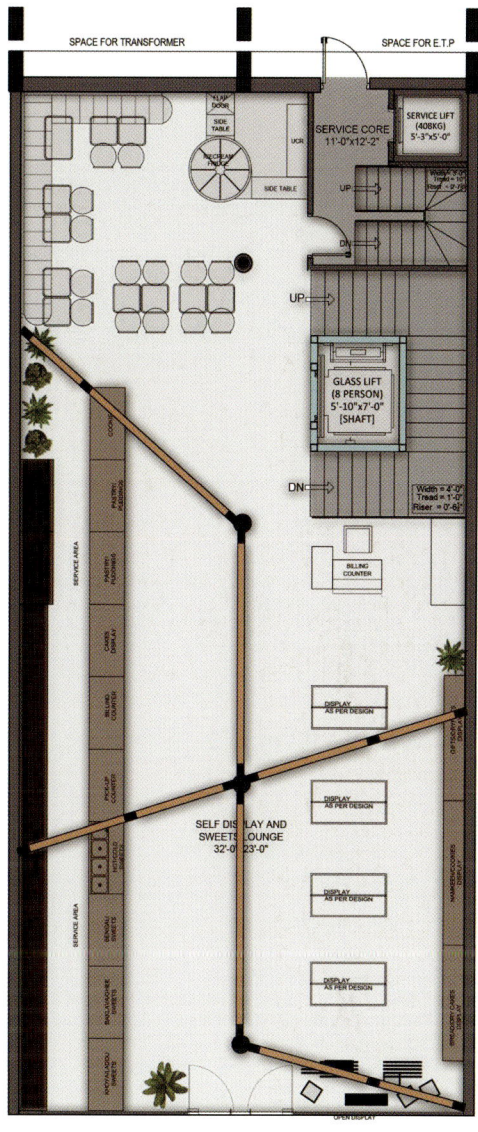

GROUND FLOOR PLAN

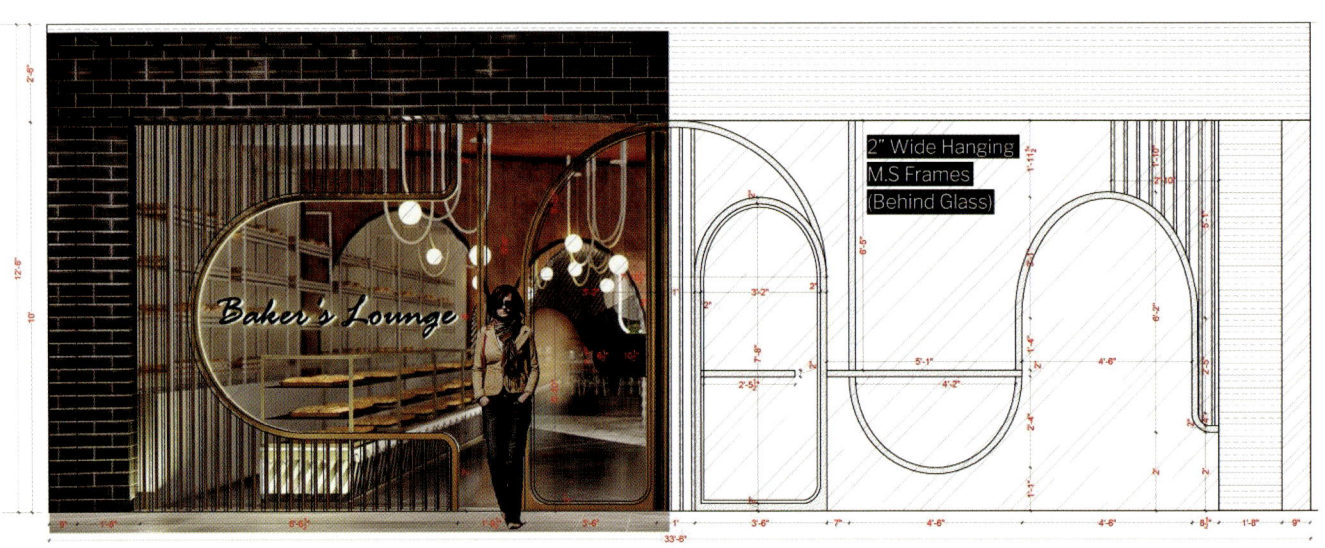

ENTRANCE GLASS DETAIL

Arches aligned at different angles to create different volume zones like service, display and sitting.

Service core planned at the rear right corner of the building adjacent to the sitting area with lift and staircase.

Transparent windows and **door** has been provided at the entrance to get the overview of display from outside.

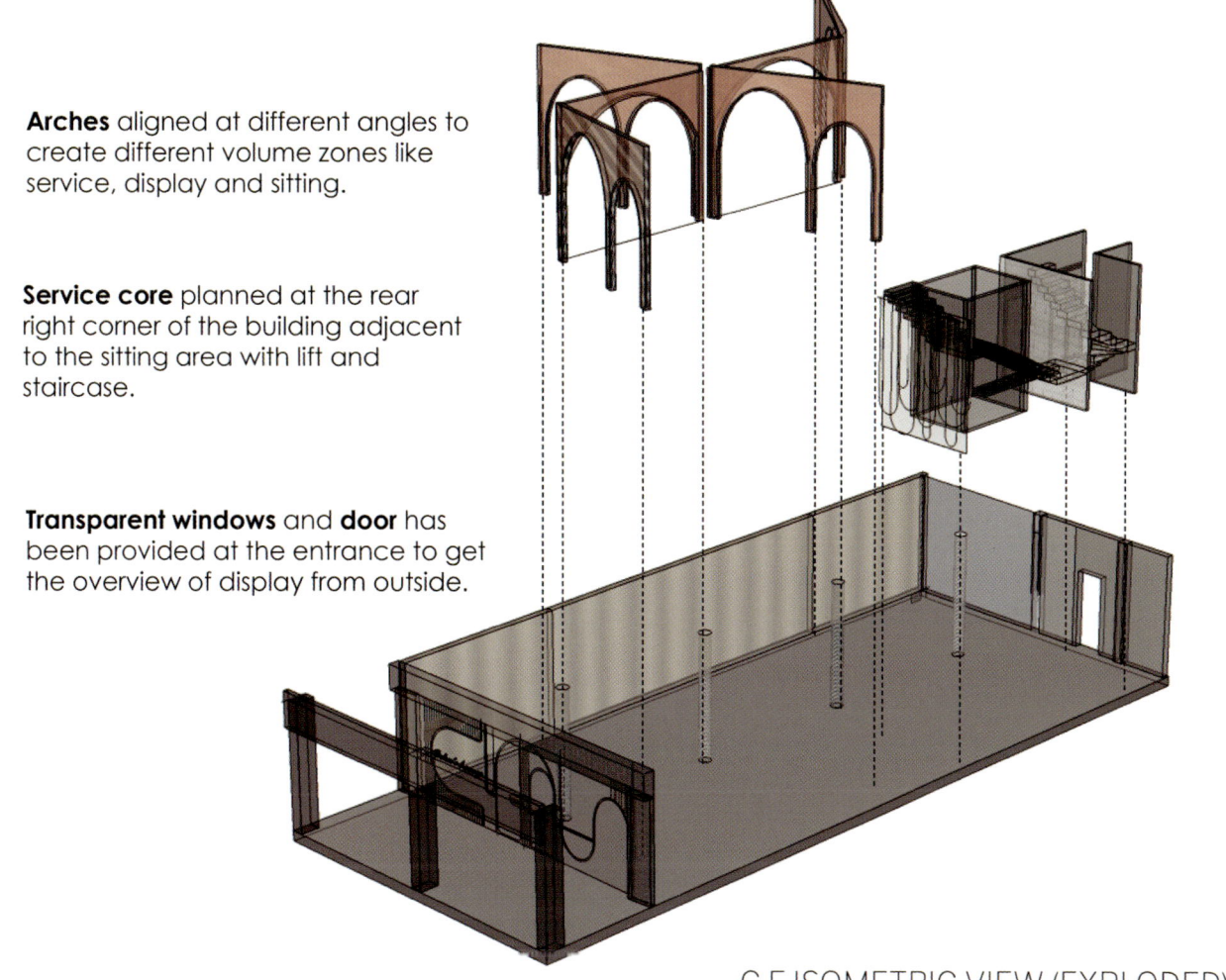

G.F ISOMETRIC VIEW (EXPLODED)

Counterstereotype visuals in Baker's Lounge are an attempt to transform the mundane and dreary shopping experience for a box of sweets. The motive was to weave a culinary experience that constantly renewed its escapade for the visitors with its divergence from the aesthetics of a regular "mithaiwala" shop.

Interestingly, the pursuit towards reimagining a habitually frequented location became an inward journey, with a quest to conceive an atmosphere that is unhackneyed. The enquiry and the process for the design of Baker's lounge, transformed into a deviation from the prejudiced assumption of a retail outlet character. The outlet continues to be the busiest sweet shop in the city, thronged by visitors at all times of the year.

Katsioulis Bakery
Athens, Greece

A Multi-collection space of taste experiences.

Branding project with logotype, corporate applications, outdoor labeling and packaging for a historical bakery – pastry in Athens southern waterfront. Strategy: Walking through its forth decade, one of the most successful Athenian brands in bakery – pastry industry broadened its vision and decided to renovate its store in full scale, converting it into a modern inclusive space, worthy of the ingenious and extra fine products that sells and serves. Within the framework of its evolution we decided to create a fresh visual communication identity, based upon the heart of the brand, preserving the consistency with the tradition that made it distinctive. We detected its heart deep inside the baking lab, which turns on the lights at two o'clock in the morning, ''before the rooster crows…!''

We are located at 4 Titanon Street in Elliniko.
2020 was the year of the pandemic, but it was also the year when our vision, perseverance and passion could not let time go to waste. So we decided to completely renovate the bakery, turning it into a modern, eclectic space with excellent flavors. In a wonderful, green and specially designed space you can enjoy your coffee, brunch, food, dessert and ice cream. Visit us and you will find that we are much more than just a bakery!

Project details

▶**Design:** SOLO
▶**Homepage:** https://studioardete.com
▶**Area:** 2265.3 ft²
▶**Location:** Titanwn, Athens, Greece
▶**Photographs:** Purnesh Dev Nikhanj
▶**Branding and graphic design:** Cursor Design Studio

About us

We are Designers and Makers.
For us, the design process starts with an idea – a pattern, a piece of furniture, a fabric innovation, an inspiration. We take that idea and craft it into a reality. We are second-generation makers, and we believe in the heritage of what we make. We use cutting edge technology for our manufacturing and if the method to fit our ideas doesn't exist, we create it. We make textiles, home-ware and hero items. Our design focus is on transformative constructions and designs. We design, manufacture and ship everything from the same building – our creative hub in Athens, Greece. Our secret weapon is our commitment to quality.

Our quality is in the details, from our exclusive fabric weaves and patterns, to our use of recycled-content material. We like to challenge ourselves, pushing for innovation in our designs and in our manufacturing.

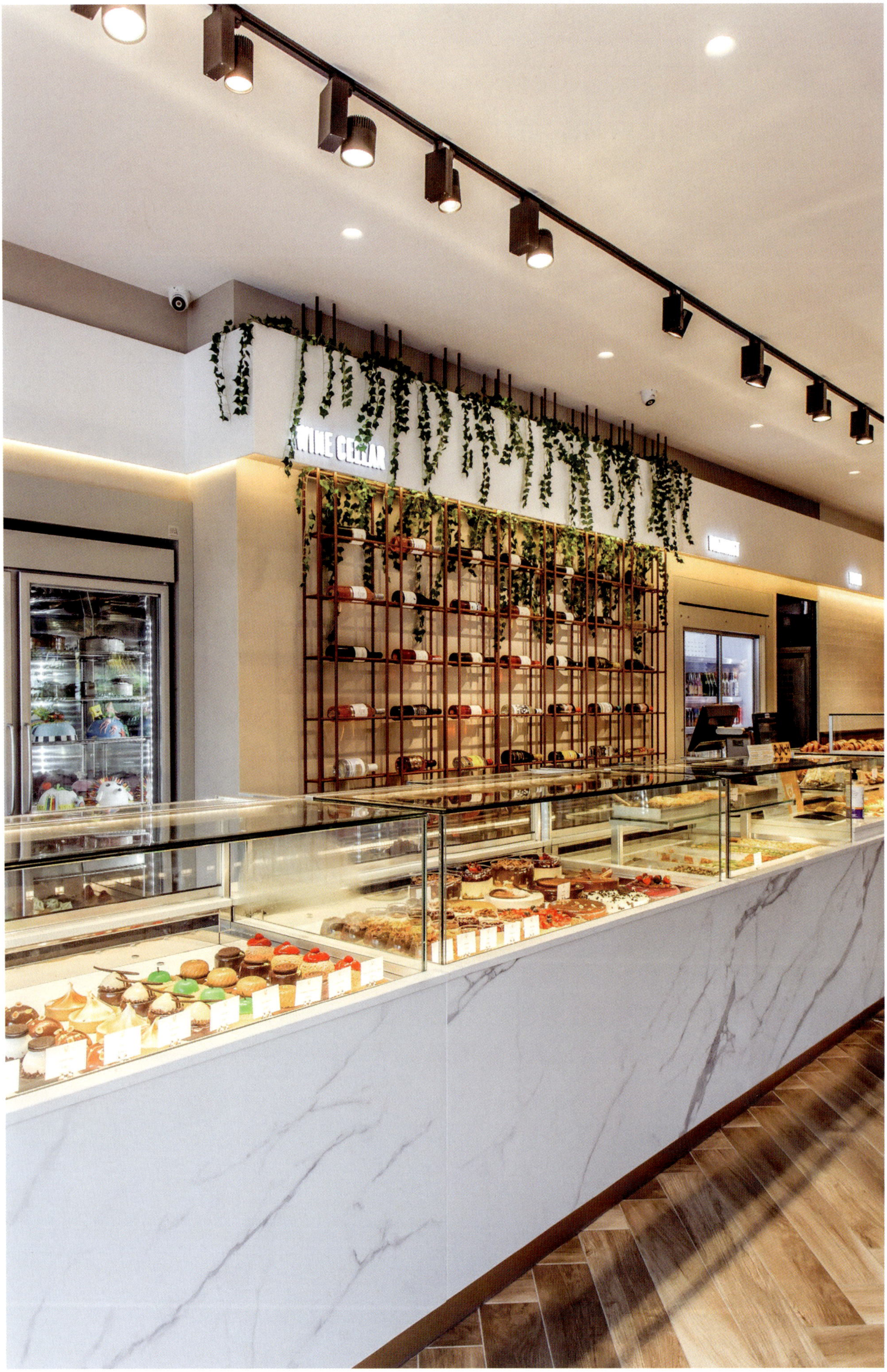
WINE CELLAR

PATISSERIE

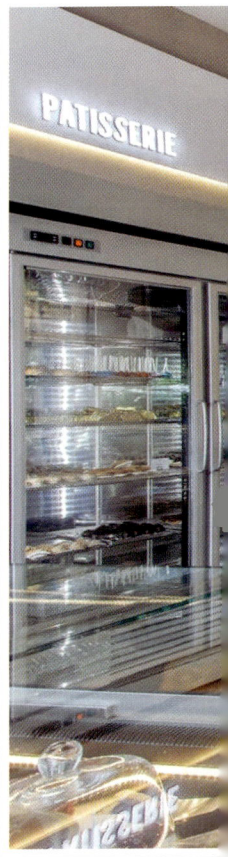

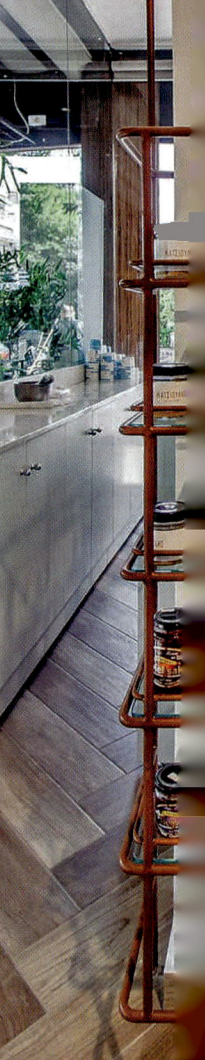

In a wonderful, green and specially designed space you can enjoy your coffee, brunch, food, dessert and ice cream.

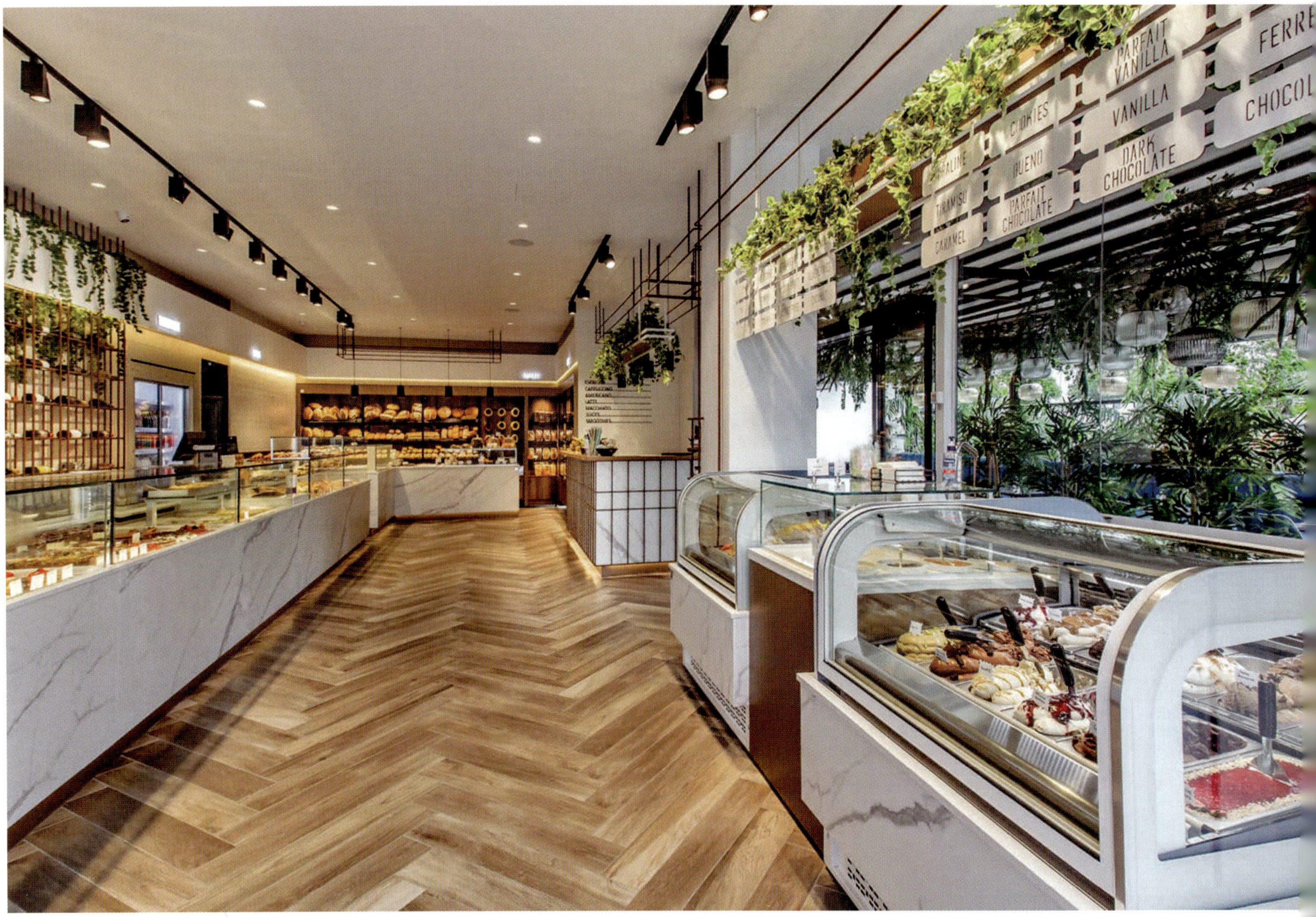

033

We decided to completely renovate the bakery, turning it into a modern, eclectic space with excellent flavors. In a wonderful, green and specially designed space you can enjoy your coffee, brunch, food, dessert and ice cream. Visit us and you will find that we are much more than just a bakery!

Voila Pastry

New Cairo, Egypt

Voila Chocolate & Pastry Shop desserts.

In November 2019, the 3rd store of the Egyptian pastry brand specializing in chocolate VOILA opened its doors in the city of Cairo. And this one is worth the detour. MMM is closely following the trend that is shaking the catering sector all over the world. The retail design is breathtaking, bordering on magic, and transports us to the history of the brand. Designed by the TAJ Design agency, it used noble materials such as marble, giving the brand an extremely chic atmosphere. The walls are covered with figurative paintings, a nod to the style of the painter Michelangelo. Mirrors, placed in strategic places, make it possible to enlarge the space by playing with the different volumes of the room. In general, the decoration particularly echoes wild nature: such as the statue in the shape of a giraffe which serves as a light fixture or the many floral installations around the shop windows and above the counters presenting the food products.

VOILA specializes in chocolate-based creations and cakes elegantly presented under glass counters. Customers can eat them on site or take them away and can even accompany them with a coffee, prepared on site. As we told you, the new pastry concepts with breathtaking retail design are particularly popular. This extensive theatricalization of the product allows the catering sector to move upmarket and also offer a unique in-store experience.

Project details

▶ **Design:** Taj Design
▶ **Homepage:** https://studioardete.com
▶ **Area:** 198.347m²
▶ **Location:** New Cairo, Egypt
▶ **Photographs:** Nour El Refai
▶ **Branding and graphic design:** Bold Branding

About us

The Best Solutions By Professional Designers. A company established by Sherif Helal in 2014; specialized in interior design, finishing, furnishing and home styling, landscape, and architectural design. The scope of work is residential, commercial, and administrative projects. It gives you the turnkey service opportunity at any project.

VOILA
Exquisite Layers

Voilà Cool Stuff – by Bold Branding Our client is a well-known desserts shop in the Egyptian market; known for its high quality, unique products with sophisticated designs and distinctive brand image. We wanted to create a sub-brand for the new pre-packed ice cream range that will match Voilà's innovative identity, remarkable packaging, with the unique design philosophy that will stand out among competitors.

Main Icon: We fashioned a unique icon for the product range to be treated as the main emblem that will be used in all the ice cream current and future range. Voilà is known for always delivering out of the box desserts, there are no boring products. The same thing with their new ice cream range. So we named the range "cool stuff" indicating the cool additions to the traditional flavours that created the Voilà twist. "Cool stuff" is hugging a nice ice cream spoon with some delicious drizzles.. surrounded by some sparks, that represents the surprising cool flavours in the range.

Design Philosophy: The challenge was to create an ice cream packaging that shows the flavours subtly without having to include the traditional fruits or ice cream images … But instead, we wanted to create an illustration language that beautifully translates the flavour as well as communicate the brand's image and the high product's quality standards. Each label has a differentiating colour combination that is relatable to the flavour; with a bold readable title in the middle, and each label includes 4 main icons; the flavour pattern, the tasting tongue, the flavour typo, and the exciting optical illusion icon

The Flavor Pattern: moving away from the traditional images and direct flavour illustration; inspired by the flavours' elements and lines; we created a distinctive geometric pattern assigned for each label; communicated the dominant flavour of the pack harmoniously with the rest of the icons and the unique colour palette assigned.

The Flavor Typo: Inspired by the flavour lines and elements we created a unique initials typo for each flavour.. connecting it with the colour palette, and flavour pattern, you will be able to easily differentiate each flavour on the shelve and quickly grab your favourite bucket!

The Tasting Tongue: Ice cream is meant to be licked, and enjoyed with all the taste buds in the tongue ... each group of flavours has a unique impact on our taste buds, thus we grouped the flavours by their taste effect into 4 main categories. (Spade) the loved classics like the all-time favourite chocolate ice cream. (Spark) the spark of the fresh taste like that of the fruity flavours (Electric bolt) the clash of flavours that creates unforgettable experiences like yoghurt berry or salted caramel. (Flower) represents all the nutty rich flavours that elevate any ice cream taste

Colour Combinations: As we were creating a new direction for the ice cream packaging we wanted the flavours to be easily differentiated in an elegant way... and choosing each flavour's colour palette was as essential as the illustrations themselves. We chose soft baby colours as the main background with more striking icons to grab the attention. Colour coding the labels was vital to differentiate the flavours and tie all the elements together to the flavour of the pack; creating an appealing ice cream package that is easily spotted on any shelf.

Jolie Sticks
chocolate

Jolie Sticks
raspberry

Jolie Sticks
pistachio

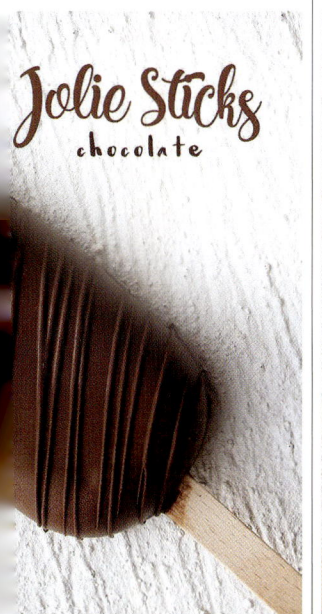

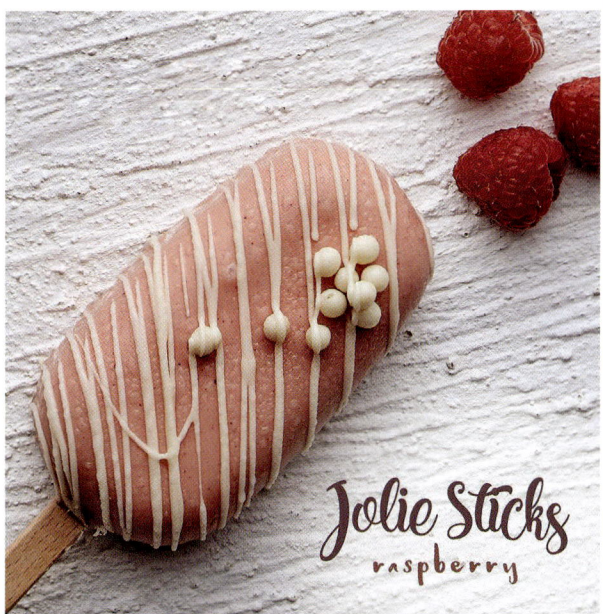

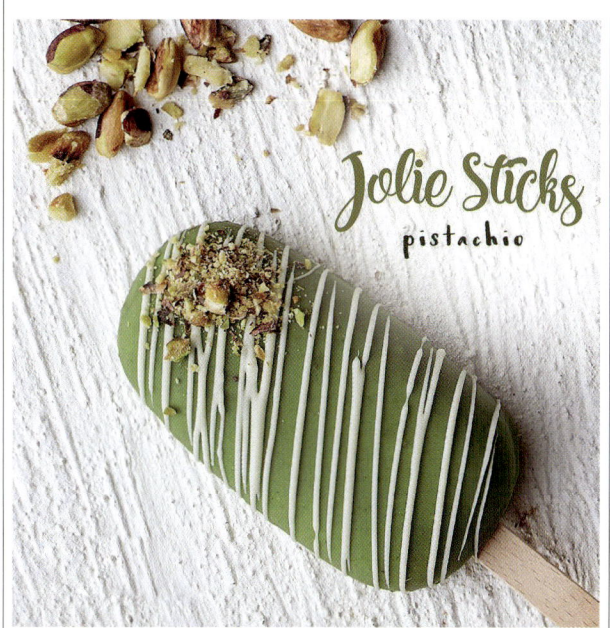

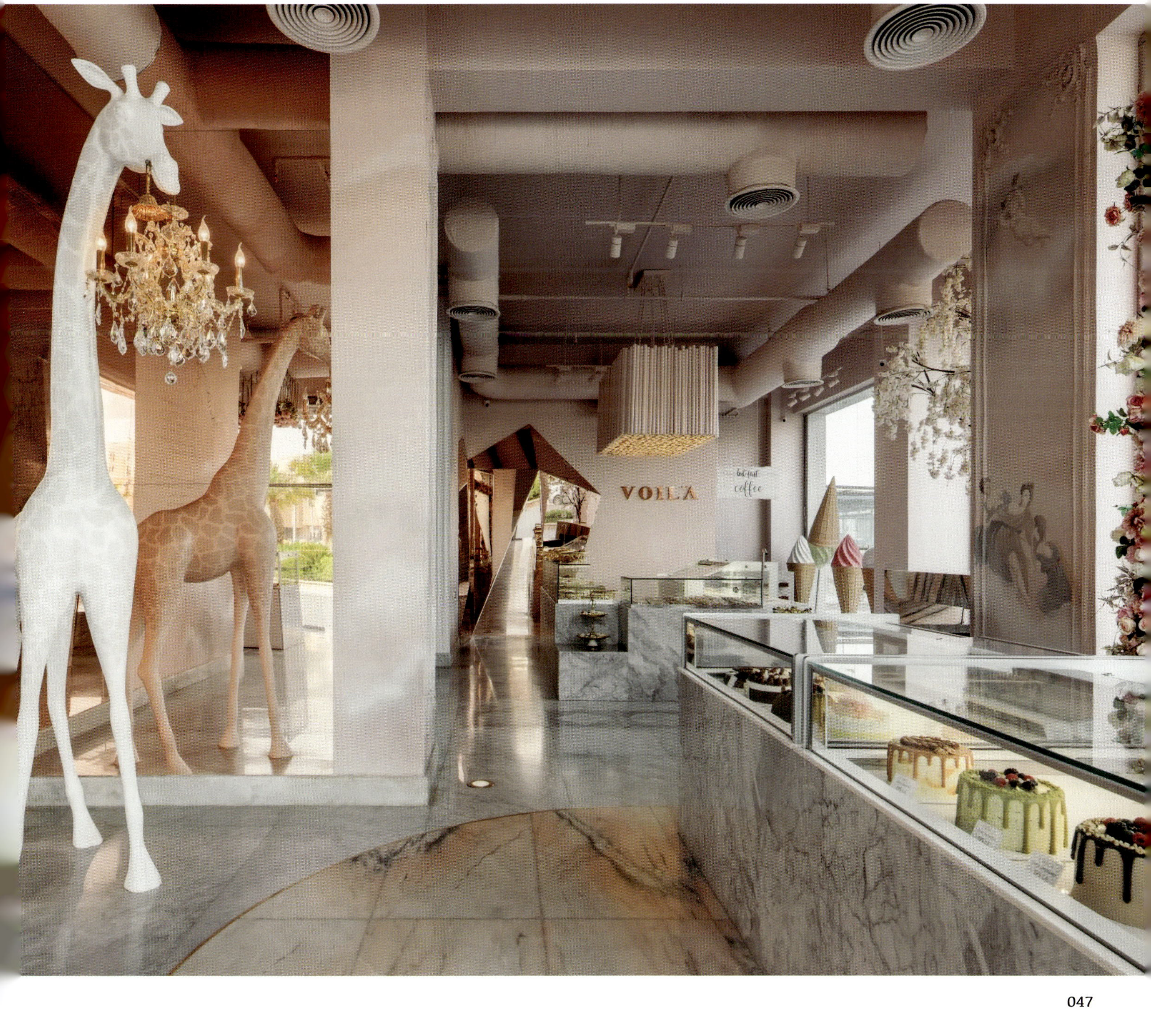
VOILA

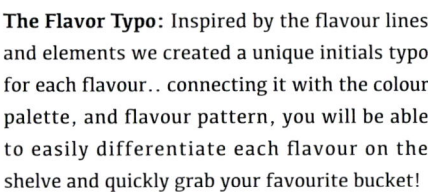
The Flavor Typo: Inspired by the flavour lines and elements we created a unique initials typo for each flavour.. connecting it with the colour palette, and flavour pattern, you will be able to easily differentiate each flavour on the shelve and quickly grab your favourite bucket!

The Exciting Optical Illusion: We all experience ice cream differently but we can all agree that ice cream brings happiness and feelings of joy. The different experiences are presented in fun optical illusion elements whether it is a dazzling, explosive, loud, oriental, soft, or fun effect.

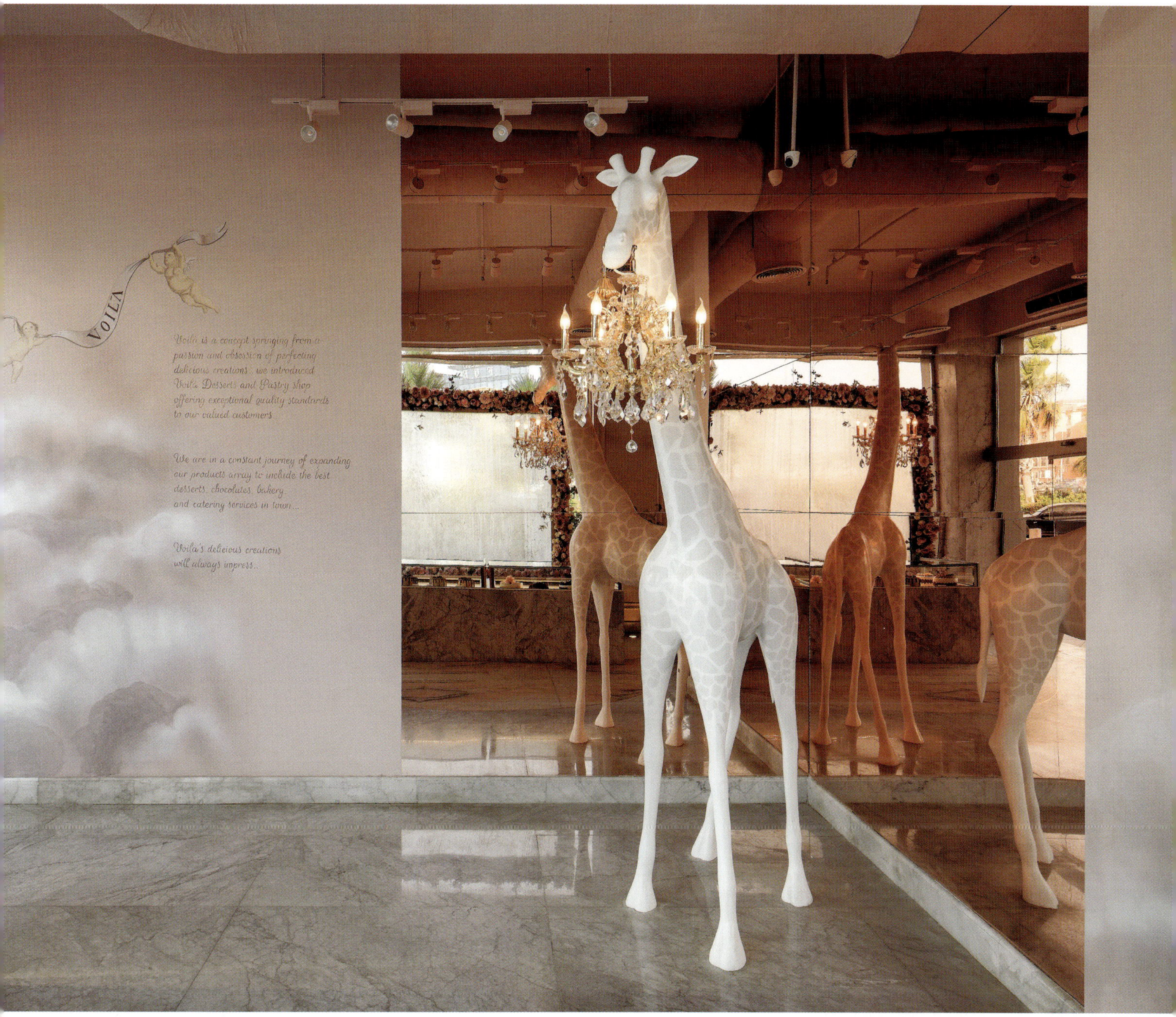

Voila is a concept springing from a passion and obsession of perfecting delicious creations, we introduced Voila Desserts and Pastry shop offering exceptional quality standards to our valued customers.

We are in a constant journey of expanding our products array to include the best desserts, chocolates, bakery and catering services in town.

Voila's delicious creations will always impress.

Flower represents all the nutty rich flavours that elevate any ice cream taste.

051

Dorbolò La Gubana, Largo Boiani, 10. 33043 Cividale del Friuli, Udine, Italy / +39 0432 727052

Udine, Italy
Dorbolò Gubana

Dorbolò La Gubana Boutique is a bakery and café.

Pastry shop, bakery, coffee spot. With its gray stone surfaces and polished brass details, the new Dorbolò concept store is a precious window overlooking the historic city center of Cividale del Friuli. A modern reinterpretation of traditional pastry shops from Mitteleuropa, the store is conceived as a small theatre set which, with an intriguing optical effect of shaped wings, directs the gaze towards the back wall dominated by the bar counter and the large brass logo. Among its hazelnut-colored wings, there are display shelves, storage spaces, and small but comfortable alcoves where you can sit down to enjoy a slice of cake and a coffee. The large display cases in the center of the space show the pastry creations in all their beauty. The Gubana, the typical dessert of this area, is certainly the main actor on stage. Bread also plays an important role with its corner placed at the entrance to show the fresh-baked products. The careful work on the interiors is completed by the lamps, all custom-made, in brass and opal glass.

Project details

- **Design:** Visual Display
- **Homepage:** https://www.visualdisplay.it
- **Area:** 198.347m²
- **Location:** Udine, Italy
- **Photographs:** Camilla Bach

About us

Visual Display is a strategy-led creative company specializing in interior design and space branding for companies and privates. Our own long expertise goes to design retails, hospitality, art or fair exhibitions, showrooms, and workspaces. With some special interior projects dedicated to private residences and villas. We manage the whole design process from the conceiving of the creative concept up to the final technical design development and the project management, as needed. No matter of location, size, or function of the space; we work every day with balanced alchemy of strong technical execution and management together with an immersive and sensorial vision and storyline.

Our mission is not only to design beautiful interiors, but places where people feel good, and fully enjoy the space that surrounds them.

We support the liberation of bread and all baked goods from all unsightly plastic bags and from humiliatingly poor ingredients. Bread is important, essential and incredibly good and good for you when taken seriously. This does not mean complicated or gimmicky. In fact, the best breads, cookies and pastries are the ancient basics, traditional recipes lovingly passed on and sometimes improved by each generation. These gorgeous traditional breads deserve to be created, sold and enjoyed in beautiful surroundings.

Our love of bread and bakeries where tradition and modern life coexist has resulted in a series of articles we headlined The Rise of the Designer Bakery. It continues to be one of our most popular and most-copied article series ever. And we are glad to add another little establishment to this series of bakeries: Dorbolò La Gubana Boutique in Cividale de Friuli, in Udine, Italy.

GUBANINE
CIOCCOLATINI CON RIPIENO DI GUBANA
150g

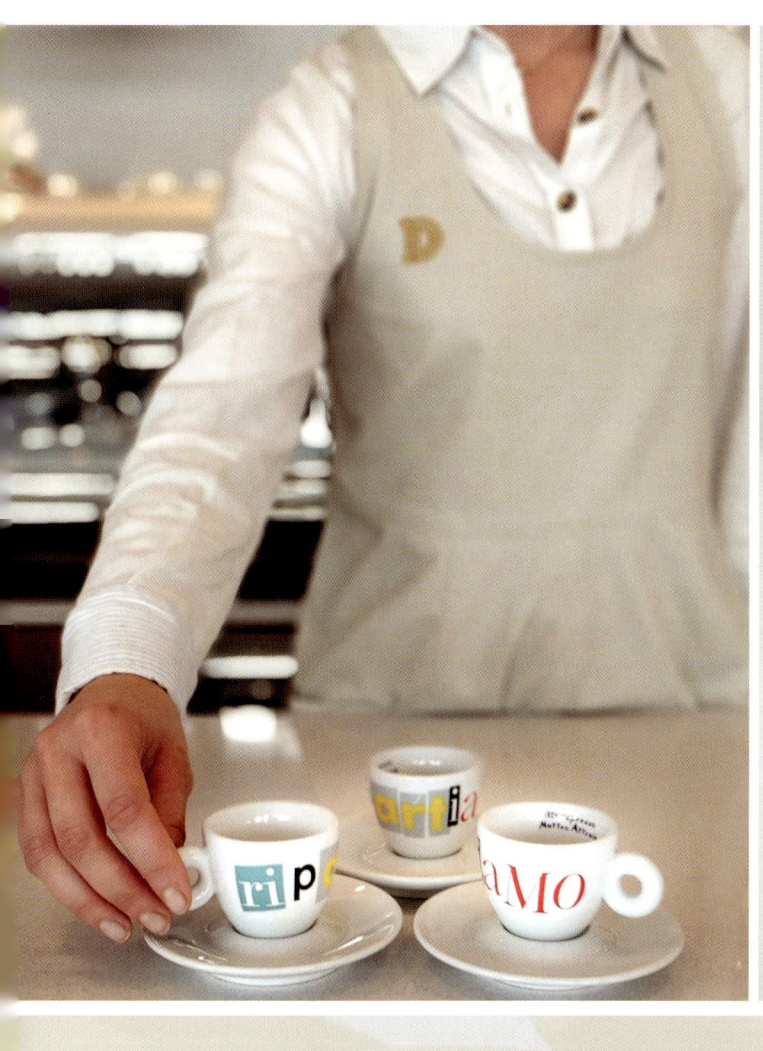

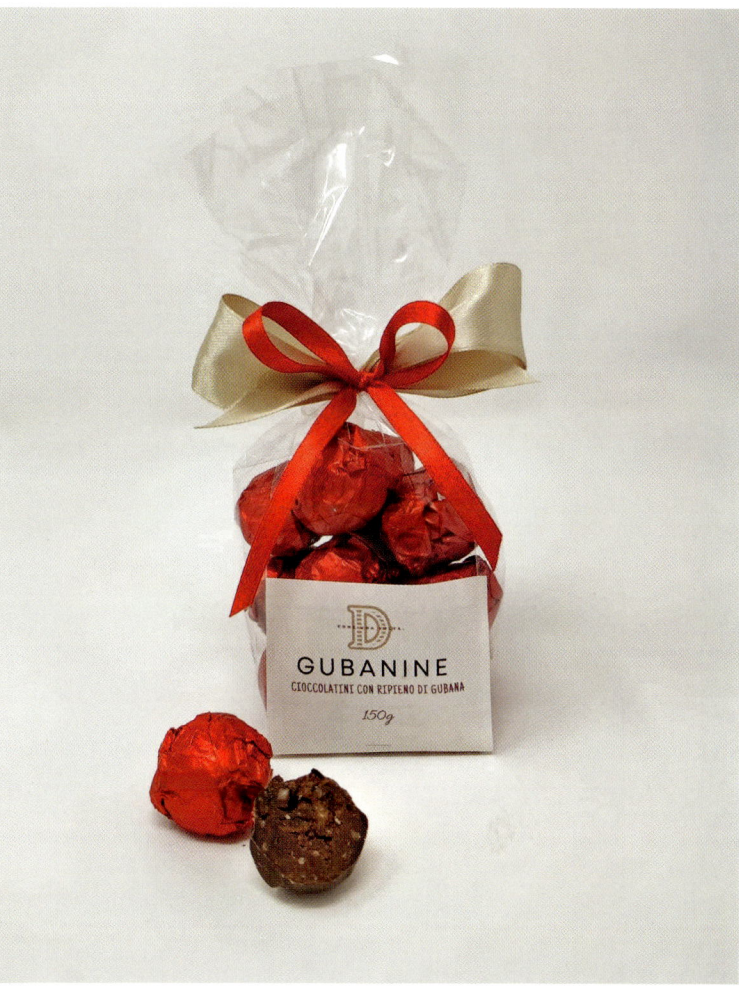

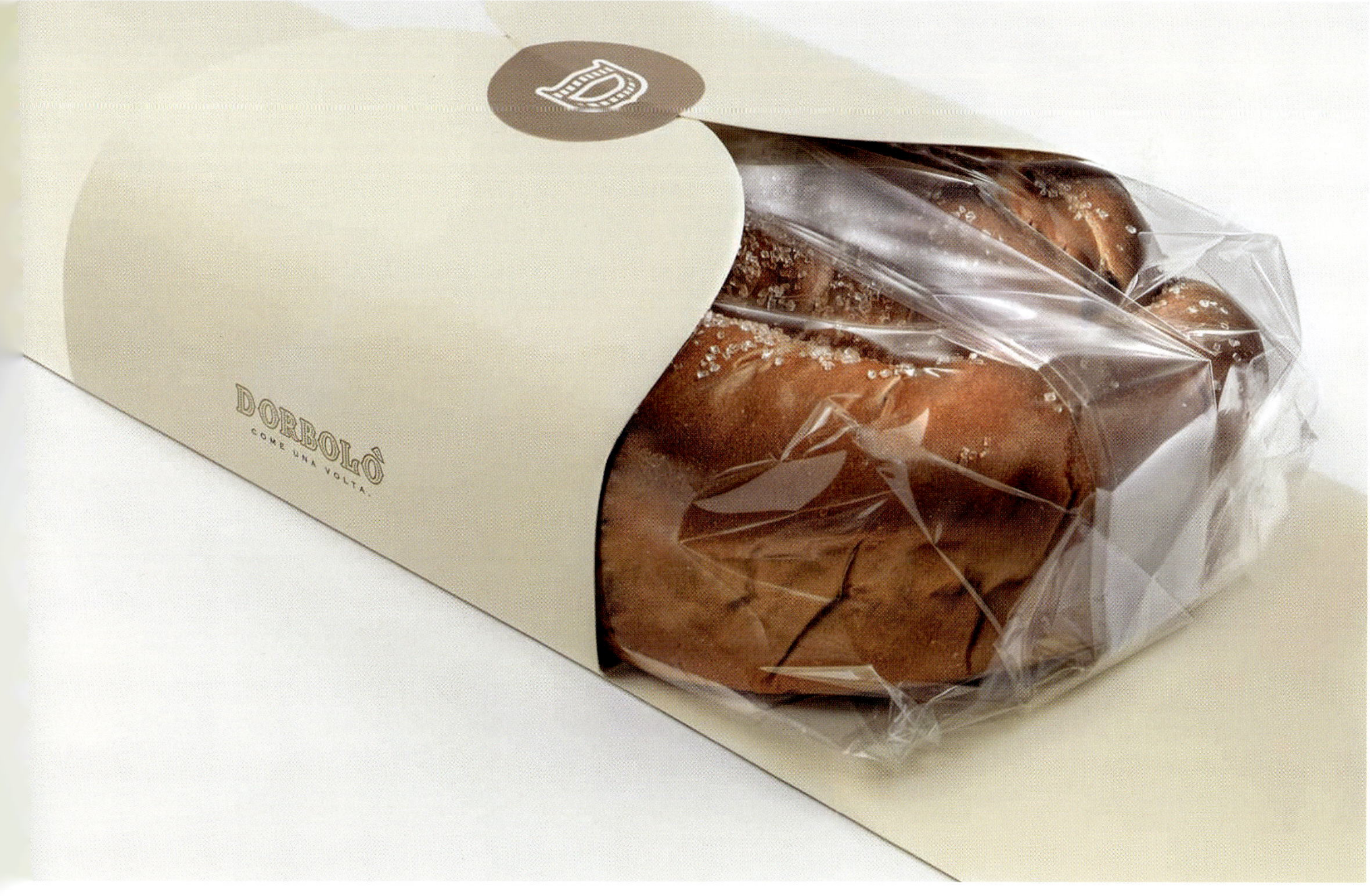
DORBOLÔ
COME UNA VOLTA

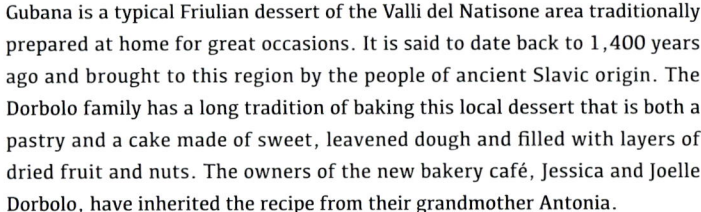

Gubana is a typical Friulian dessert of the Valli del Natisone area traditionally prepared at home for great occasions. It is said to date back to 1,400 years ago and brought to this region by the people of ancient Slavic origin. The Dorbolo family has a long tradition of baking this local dessert that is both a pastry and a cake made of sweet, leavened dough and filled with layers of dried fruit and nuts. The owners of the new bakery café, Jessica and Joelle Dorbolo, have inherited the recipe from their grandmother Antonia.

The majority of the space is dedicated to the marble display cases, but in the background, there are small round tables with intriguing L-shaped brass legs that bolt into the wall-mounted seating.

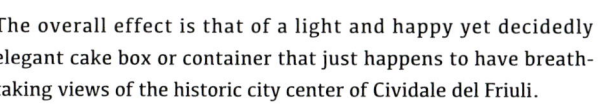

The overall effect is that of a light and happy yet decidedly elegant cake box or container that just happens to have breathtaking views of the historic city center of Cividale del Friuli.

In addition to gubana, the shop sells other cakes and pastries and traditional breads. Its interior is minimalist in all other aspects except in the ceiling that has an arched substructure that could be said to resemble the layers of a gubana when it is cut into slices. The designer tells us that the shop is envisioned to be a small stage with its various sets up in the wings. The majority of the space is dedicated to the marble display cases, but in the background, there are small round tables with intriguing L-shaped brass legs that bolt into the wall-mounted seating.

Panistas Bakery: Hernan Cortes 42, Santander, Spain / +34 641 12 32 03

Panistas Bakery
Santander, Spain

If you don't settle for the usual bread either, this is your place.

Panistas is the new bakery located in the center of Santander. The client shows us their enthusiasm to make a bakery different from the others, both in product and aesthetics. For this reason, in this project, the creative process does not revolve around the bread and its imaginary, but around its origin. With this idea, we create a beautiful, warm, and pleasant atmosphere, a new environment for the buyer. The bread and the products will complete the space giving the connotations of a bakery.

We go back to the essence, to the earth, to that first step before being cereal, later flour, and finally bread. We want to use the earth both as a material and as a chromatic range. In our research, we have a range of products made 100% with the earth that allows us to provide our space with those textures and colors that transport us to the origin. The earth is an inspiring and determining element in the choice of textures and color palette. In this search for the origin and essence, we chose to complete the project with the use of natural materials found in their raw state in nature. The original structure of the premises is uncovered, leaving beams and pillars with natural wood. The earth block, made of raw material in different formats, is the protagonist material in the project, being present in the main elements that configure the space. Black iron, on the other hand, is used to frame the latter giving us the artisan and productive character.

Project details

▶**Design:** Zooco Estudio
▶**Homepage:** https://zooco.es
▶**Area:** 132m²
▶**Location:** Santander, Spain
▶**Photographs:** Imagen Subliminal (Miguel de Guzmán + Rocío Romero)

About us

Zooco is an architectural studio founded in 2009 by Miguel Crespo Picot, Javier Guzmán Benito and Sixto Martín Martínez, architects by the School of Architecture of Madrid. The study covers a wide spectrum from the design perspective. From large-scale building to furniture design. The adapted solutions to the client and their needs, and the use of a timeless language, are the hallmarks of a studio in permanent search of new challenges.

Currently, we have offices in Madrid and Santander, where we develop projects both nationally and internationally. The work during these ten years covers more than 50 works published in national and international specialized media, some of which have been recognized in various awards such as the prestigious Design Vanguard 2019, NAN Construction 2019 + 2020, Frame Awards 2020, or the European 40 under 40 2020. Recently, Zooco has been chosen by Forbes magazine as one of the 100 most creative Spanish people in the business world.

▶**Contacts** e-mail: info@zooco.es / Call us: +34 917 355 490

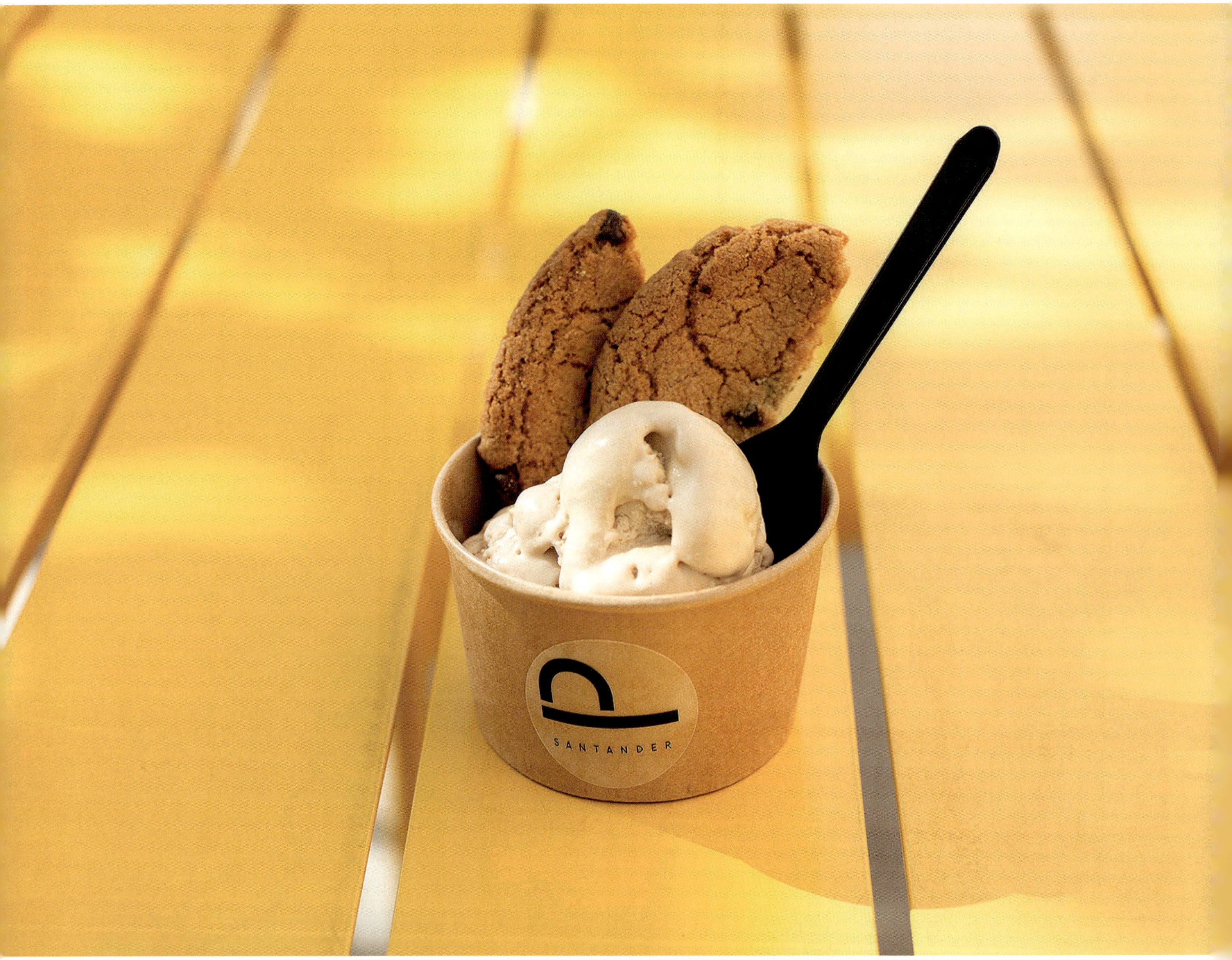
SANTANDER

Navidad en
PANISTAS

SOBAO
PASIEGO
100% MANTEQUILLA
140g

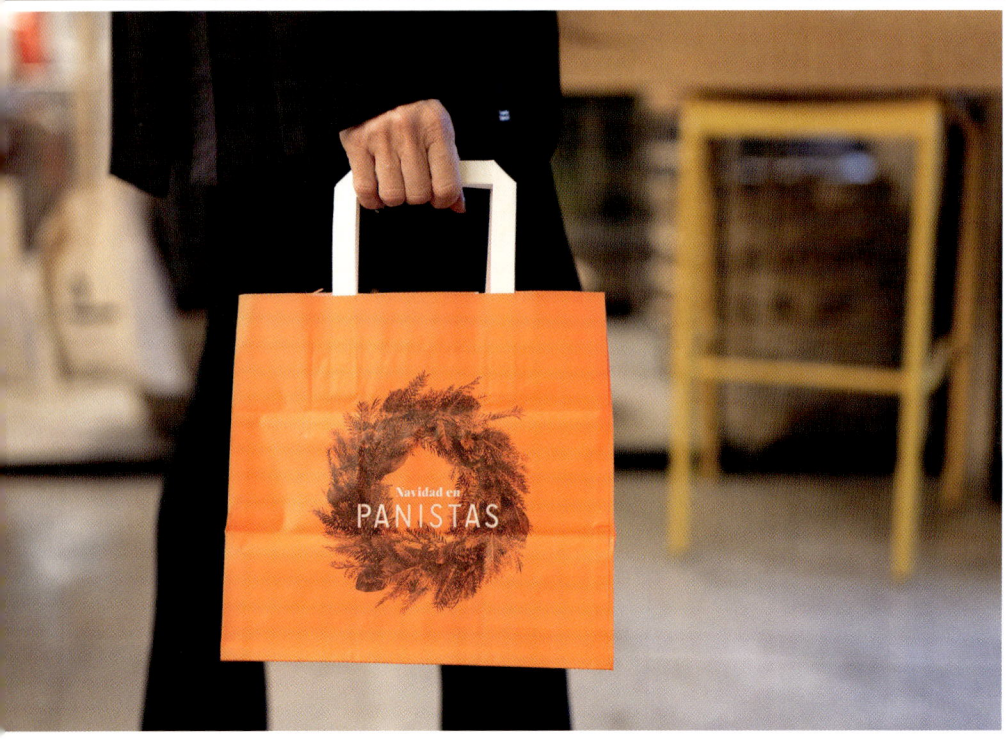

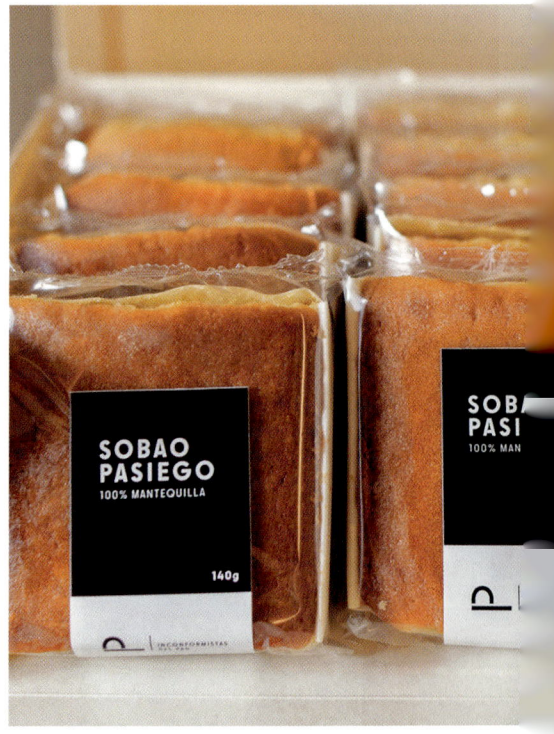

The distribution of the space consists of a commercial area for the sale of bread and related products, and a workshop as a production space. Although they are independent spaces, a visual relationship is pursued that allows customers to observe the production of bread and note its artisanal character. In the commercial area, the space is organized around the counter piece, which can be transformed into a bench to serve tables, where the consumption area opens to the outside through the wide openings in the façade. On the other hand, behind the counter, there is a display cabinet, which houses the presentation of bread and multiple services. The circulations are fluid and allow a practical relationship between the different uses of the store, rethinking and defining a new concept of a bakery.

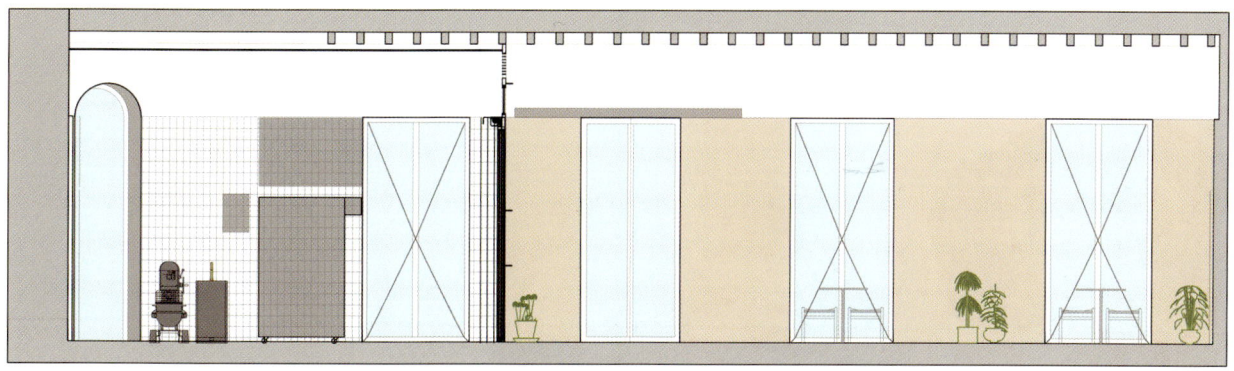

Section Plan - a

Floor Plan

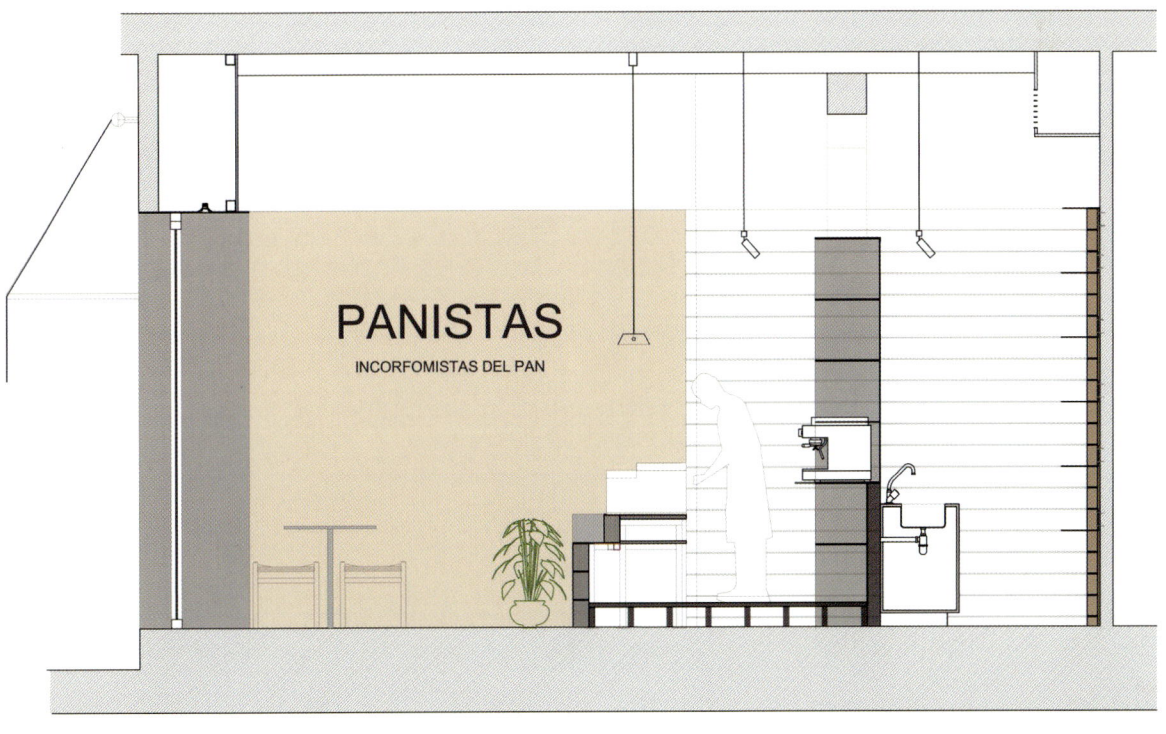

Section Plan - b

Il Punto Gelato

Livorno, Italy

An ice cream shop with a unique and unmistakable style, just like the flavors they manage to create.

▶ Il Punto Gelato: piazza damiano chiesa 9, Livorno, Italy / +39 349 879 5794

An ice cream parlor with a unique and unmistakable style, just like the flavors they manage to create. Grits that recall the decorations of ice cream, soft and playful architectural shapes are the concepts behind this project. The first of a series of shops of the "Punto Gelato" franchise.

An ice cream shop with a unique and unmistakable style, just like the flavors they manage to create. Grits reminiscent of ice cream decorations, soft and playful architectural shapes are the basis of the project concept. The first of a series of shops of the "Punto Gelato" Franchising.

Project details

▶**Design:** MODO architecture + design

▶**Homepage:** https://www.modoarchitettura.com

▶**Area:** 98m²

▶**Location:** Livorno, Italy

▶**Image:** MODO architecture + design

About us

MODO architettura + design is an established architectural and interior design studio based in Livorno and Milano, created through a collaboration between Sondra Pantani and Pietro Marsili, both graduates in Architecture at the University of Florence.

The Studio deals with all aspects of Architecture and Interior Design, covering each and every aspect of the project: from concept to completion. We exercise our creativity and passion for architectural and interior design introducing new and exciting ideas into a wide range of residential, commercial, hospitality and workplace spaces. We pay special attention to the details and materials of each project, making sure that our work is original, unique and fitting for each client. We do this by also respecting the surrounding environment and the singularity of each space.

▶Contacts e-mail: info@modoarchitettura.com / Call us: +339 8351073

IL PUNT•GELATO

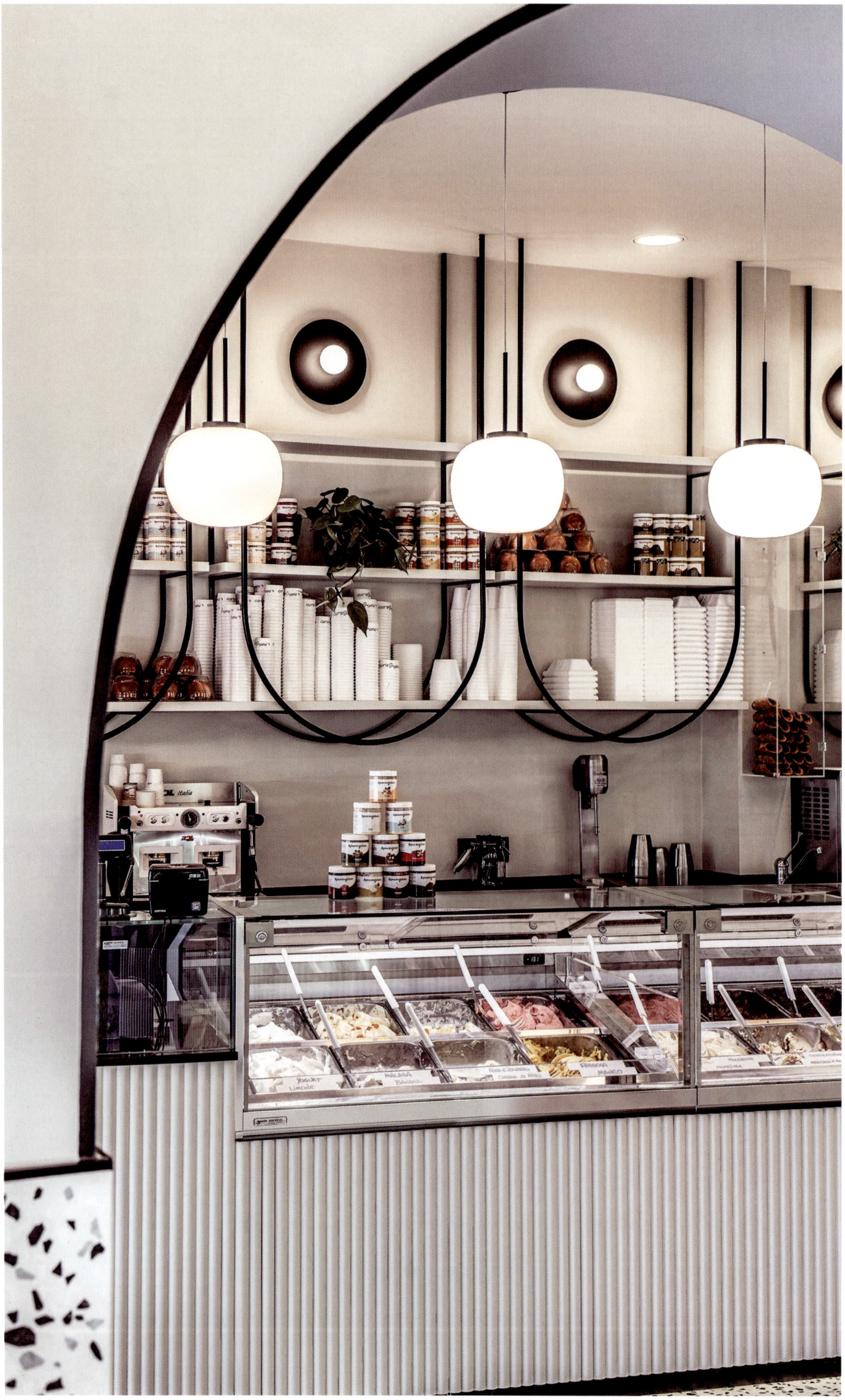

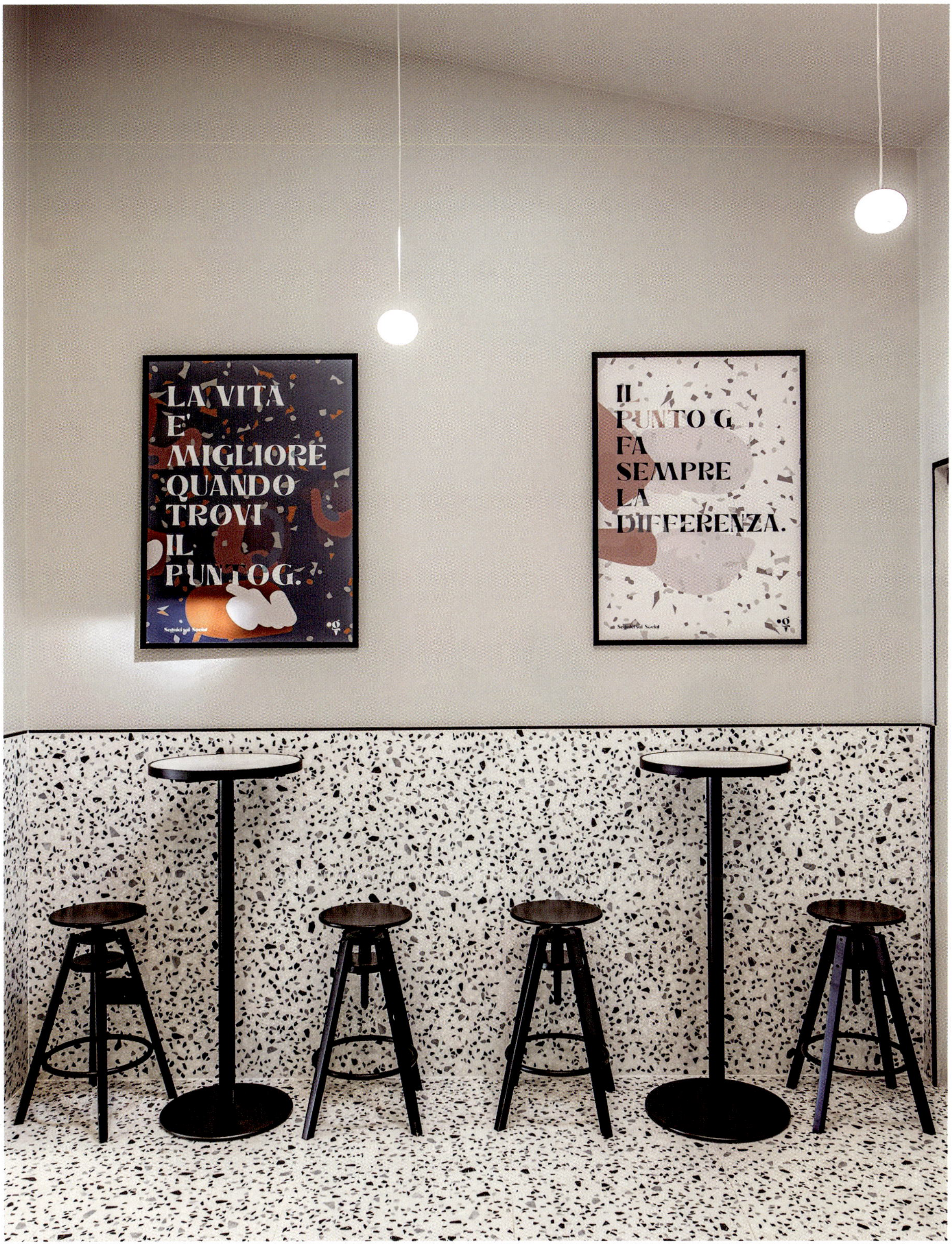
LA VITA
E'
MIGLIORE
QUANDO
TROVI
IL
PUNTO G.
IL
PUNTO G
FA
SEMPRE
LA
DIFFERENZA.

il punt•gelato
Fondente
il punt•gelato
Giotto
il punt•gelato
Pistacchio
il punt•gelato
Arachide
il punt•gelato
Classica

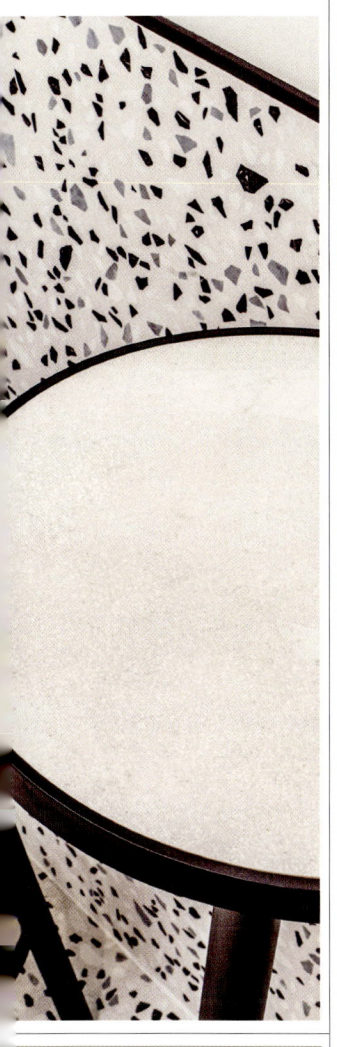

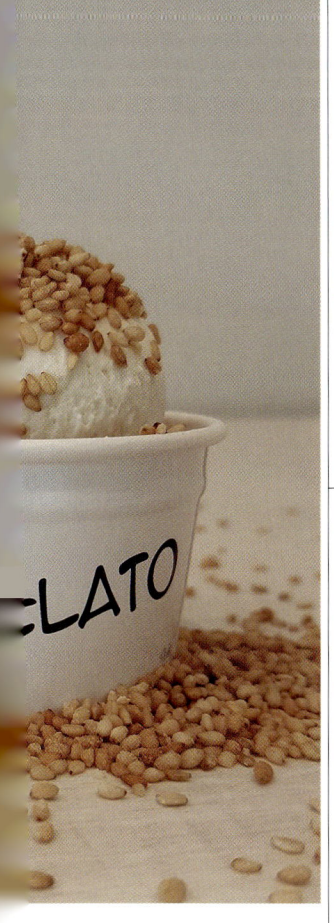

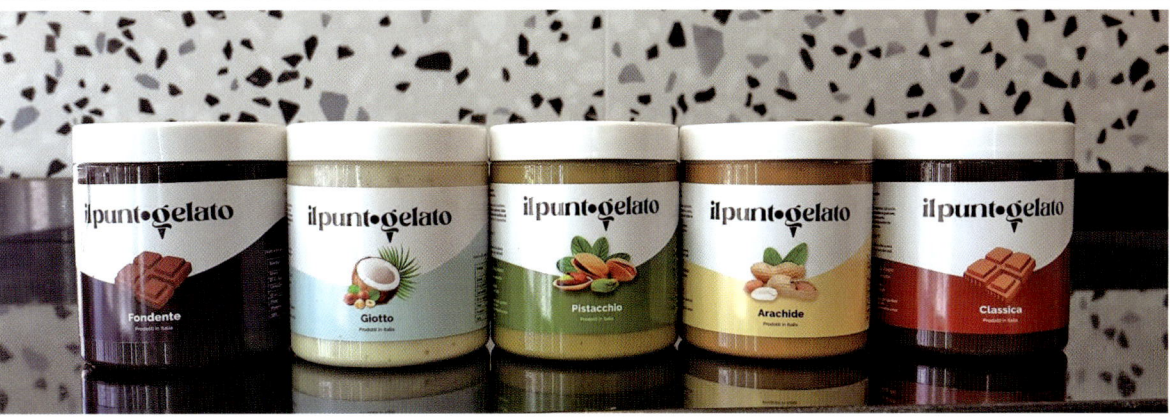

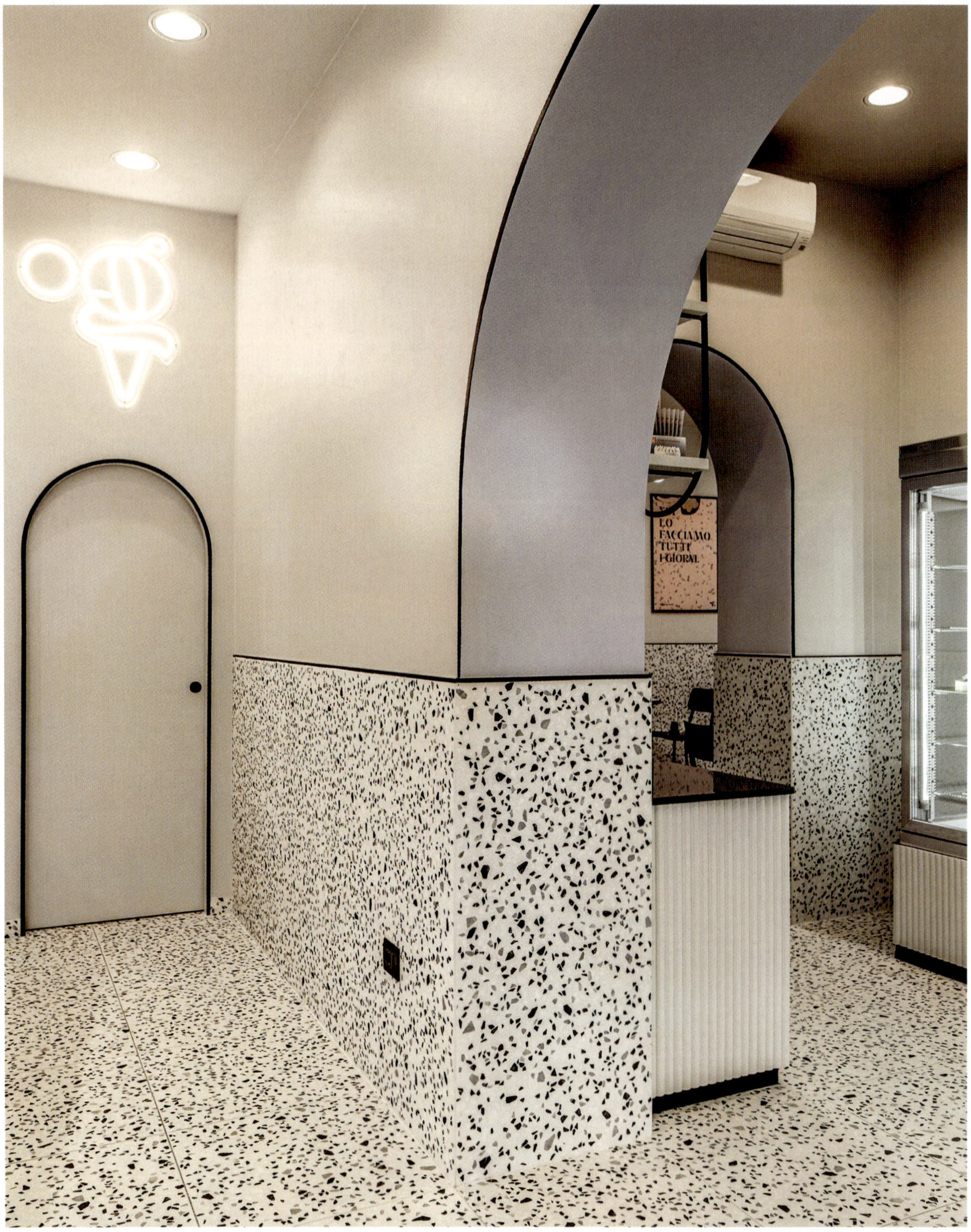
LO
FACCIAMO
TUTTI
I GIORNI

ANNA & SARA
BUON
COMPLEANNO

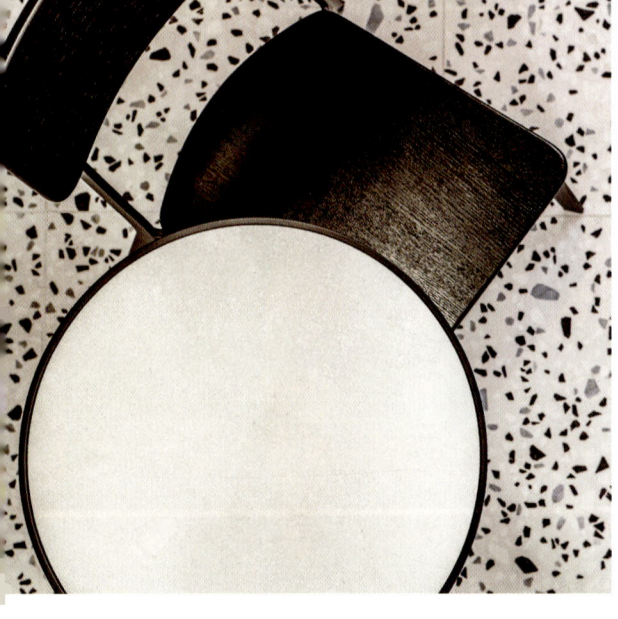

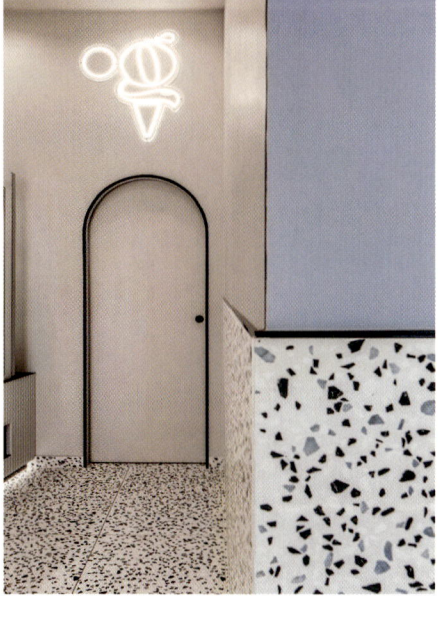

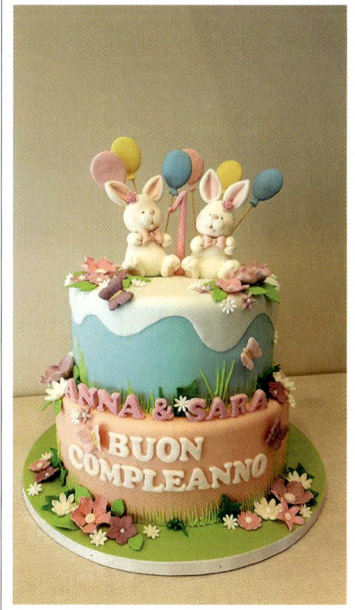

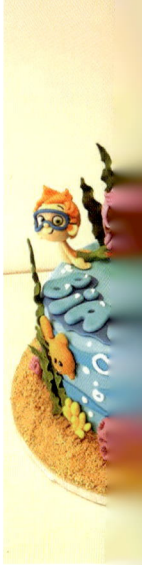

In the first point of sale of the Punto Gelato franchise in Livorno, the design gives the place a unique style, where each environment has been designed for a different function, from take-away consumption to the small room for those who want to consume at the table. Leitmotiv, the grit tiles, which recall stracciatella ice cream

Sacher Eck Wien: Kärntner Str. 38, 1010 Wien, Austria / +43 1 514560

Sacher Eck Wien
Wien, Austria
Where tradition meets the future, Classically Viennese newly composed - Confectionery & Cafe

For what is probably the most famous cake in the world, the Original Sacher Torte, it was to create an appropriate ambience – from this claim the finest corner of Vienna, the new Sacher Eck, was achieved. The famous Hotel Sacher, in its prime location opposite the Vienna State Opera, is a long-established family business that exudes elegance and charm. It unites classic Viennese design with its heritage as a world-renowned Austrian hotel and restaurant brand.

The decision to refurbish the Sacher Eck was preceded by a long maturing process in which the scope and the design orientation were very carefully considered. The hotel and restaurant projects of the Vienna-based architecture firm BWM Architekten under the direction of interior design expert Erich Bernard have attracted a lot of attention in recent years. One such project was the conversion of the traditional Viennese jeweller A.E. Köchert, which involved mordenising and reinterpreting Theophil Hansen's historic premises. It is precisely this specific interior design expertise that won the firm the contract for the redesign of the Sacher Eck. The guideline for the redesign of Sacher Eck was laid down in close coordination with the owner family. Building on Sacher's great tradition, contemporary design elements and a modern infrastructure serve to stimulate the discourse between old and new.

Striking design elements, such as the monumental chandelier, are effectively showcased by the extension of the café and confectionery and the new ceiling opening connecting the ground floor with the mezzanine.

Project details

▶**Design:** BWM Architekten
▶**Homepage:** https://bwm.at
▶**Area:** 110 m²
▶**Location:** Wien, Austria
▶**Photo:** Christoph Panzer

About us

BWM Architekten are a multi-national architecture office that operates throughout Europe. Their main areas of focus are architecture, interior design, culture and hospitality. Founded in 2004 and led by Erich Bernard, Daniela Walten, Johann Moser, Markus Kaplan and András Klopfer, the company and the 70-member team stand for a personal approach and a cooperative development process. Whether designing interior spaces, residential and urban construction projects, or museum and exhibition concepts, BWM Architekten always work out a specific project's unique formal language and the corresponding design concept in strategic workshops with the client.

Some standout projects in the area of exhibition design are the Austrian EXPO pavilion in Astana in 2017 (which won the distinguished Red Dot Award), the Austrian Literature Museum and the hdgö House of Austrian History (German Design Award 2020: "Excellent Architecture – Winner", Iconic Awards 2020: „Innovative Architecture – Selection ") as well as the new design for a visitor's centre at the Vienna State Opera.

SACHER

The new design draws on traditional, imperial Viennese interiors, of which the iconic Hotel Sacher is a prime example. The classic Sacher colours – wine-red, gold and black – and the typical Viennese materials of velvet, brass, dark wood and black and white marble merge to form new compositions. Striking design elements, such as the monumental chandelier, are effectively showcased by the extension of the café and confectionery and the new ceiling opening connecting the ground floor with the mezzanine. In addition to the pre-existing confectionery ("Sacher Confiserie"), where original Sacher products can also be purchased, there is now a large new café area on the bel étage with a view onto Kärntner Strasse and the Vienna Opera House.

The monumental two-storey chandelier, an eye-catching centrepiece created by Viennese light designer Megumi, breaks through the ceiling to the mezzanine and creates a visual connection between the two floors. Sacher Eck's new look is largely characterised by the new spatial arrangement of the shop and confectionery with self-service area. The lighting concept, together with the new domestic engineering, adds to its visibly modernised appearance. This also gives rise to new forms of display design: The story of the Original Sacher-Torte is told through various original objects placed between the products on the shelves, under cake cloches and in small showcases, as well as through contemporary media such as the presentation developed in cooperation with the Ars Electronica Festival.

Mauve, red and gold on the mezzanine

Original Sacher products are offered for sale in a white marble room. As a material typically used in confectioners' kitchens, the white marble with its brass trim references not only the classic Viennese café atmosphere but also the world-famous Sacher-Torte, meticulously crafted by hand to this day. In contrast, the predominant material in the café area is black marble, which, together with the red velvet furnishings, creates a warm, intimate atmosphere. A bar forms the elegant conclusion of the room, while the mirrored back wall behind it creates a sense of continuity. The marble flooring with panelled diamond shapes in black and white and the matching wooden ceiling are the unifying elements that marry the two contrasting parts of the room.

The monumental two-storey chandelier, an eye-catching centrepiece created by Viennese light designer Megumi, breaks through the ceiling to the mezzanine and creates a visual connection between the two floors. Sacher Eck's new look is largely characterised by the new spatial arrangement of the shop and confectionery with self-service area. The lighting concept, together with the new domestic engineering, adds to its visibly modernised appearance. This also gives rise to new forms of display design: The story of the Original Sacher-Torte is told through various original objects placed between the products on the shelves, under cake cloches and in small showcases, as well as through contemporary media such as the presentation developed in cooperation with the Ars Electronica Festival.

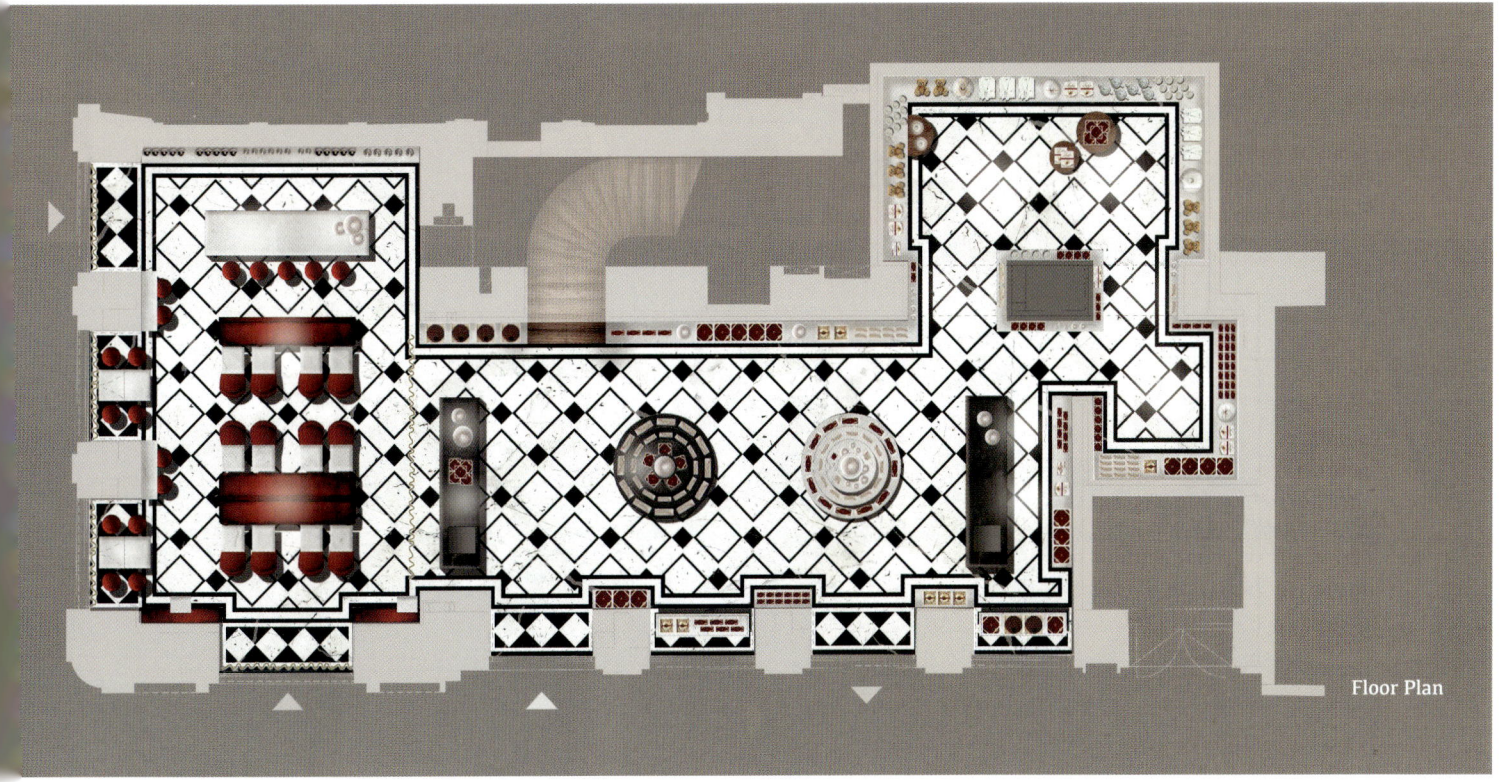

Floor Plan

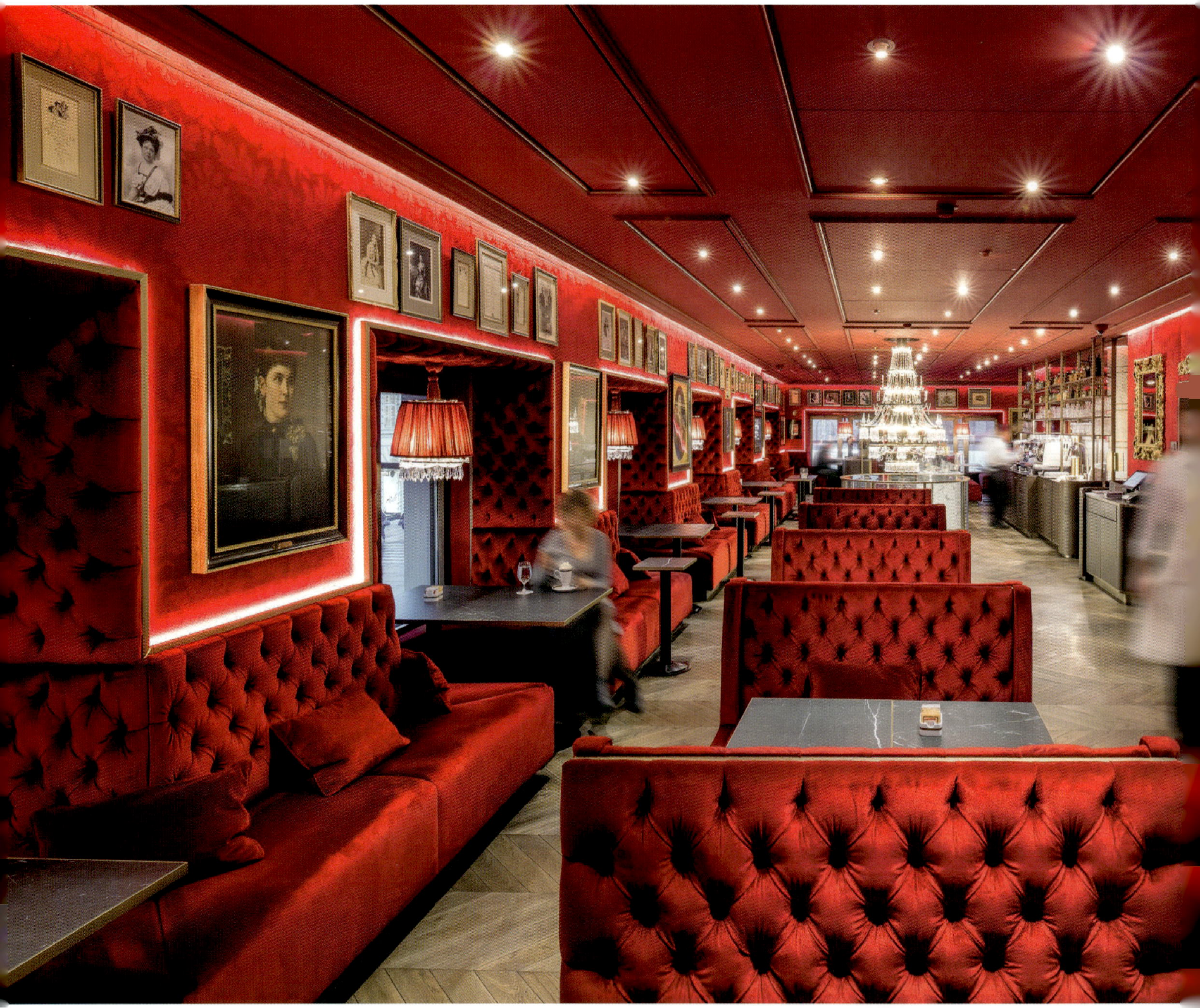

SACHER.
DAS ORIGINAL.
SEIT 1832.
SACHER

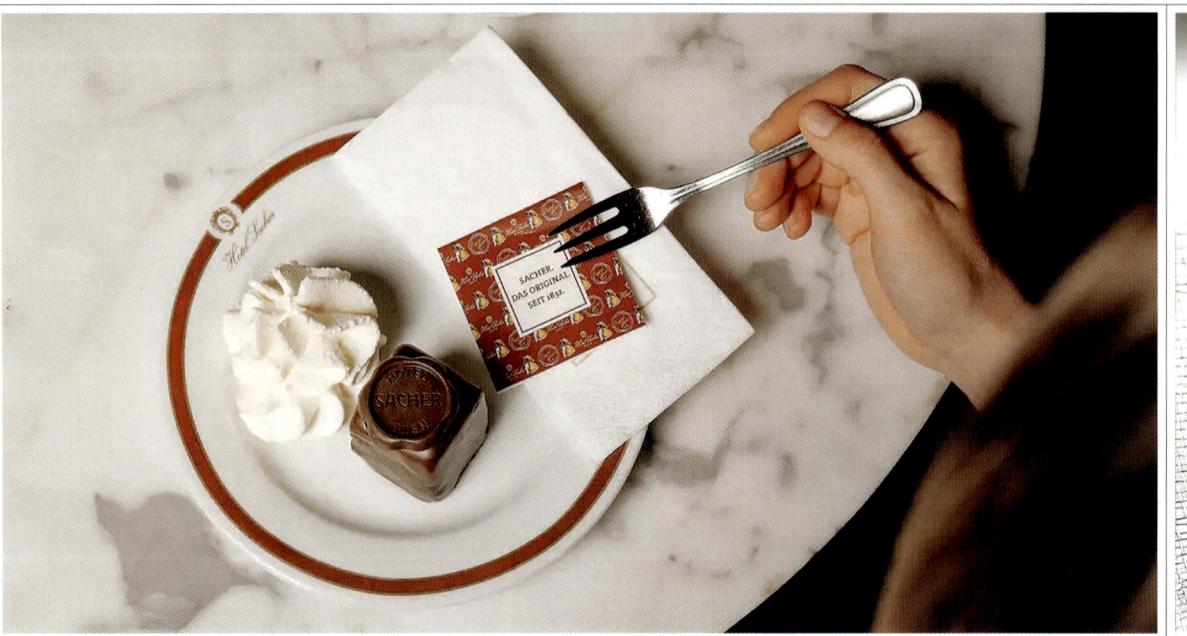

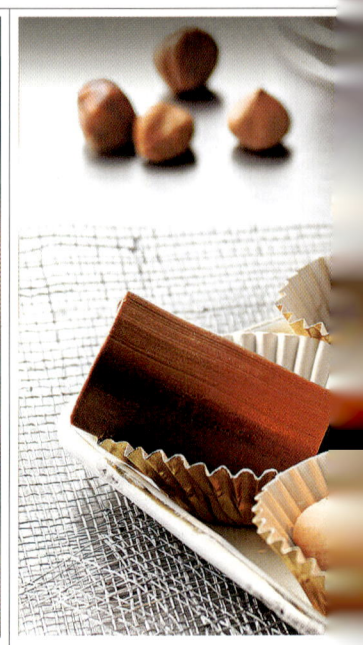

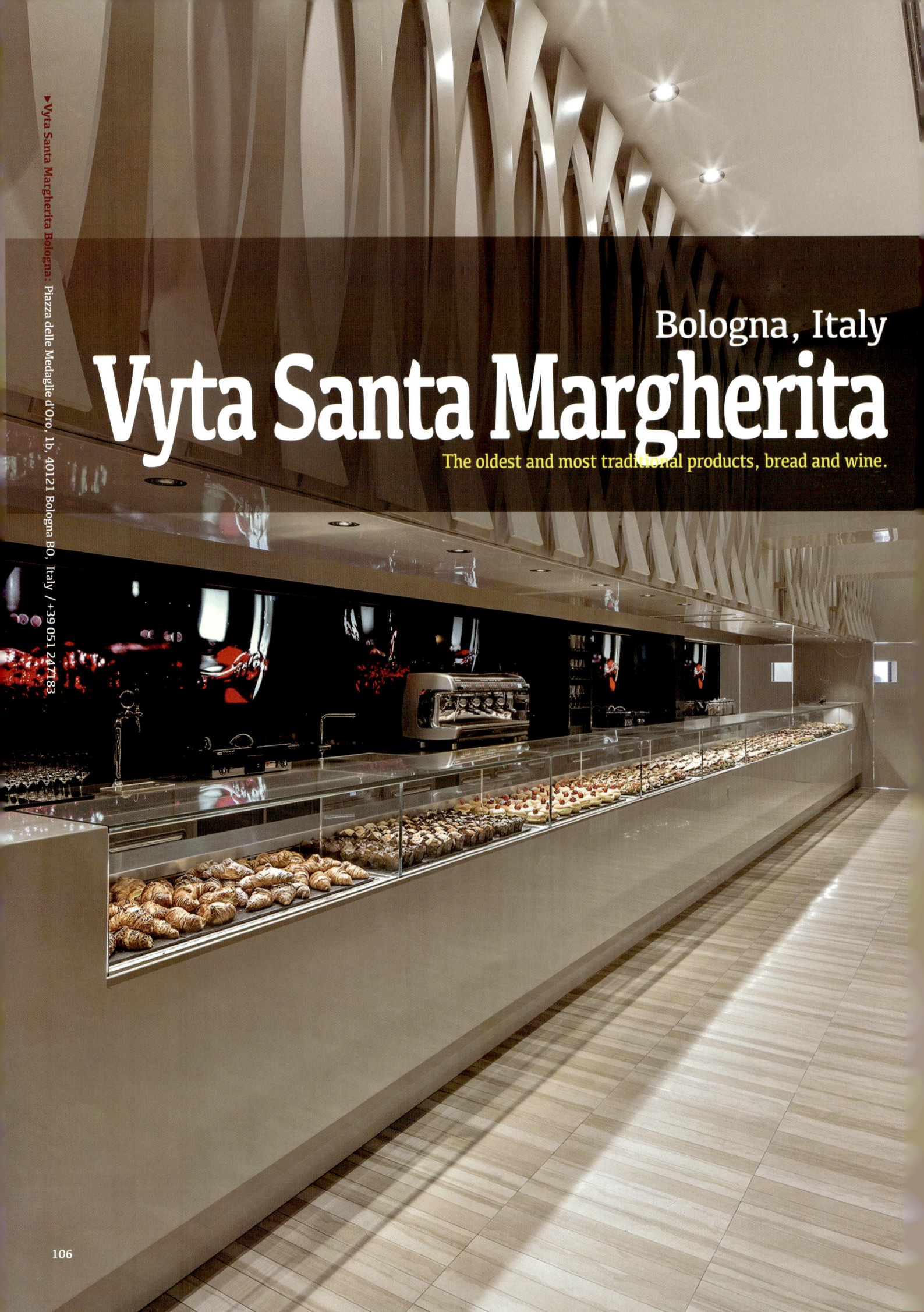

Vyta Santa Margherita
Bologna, Italy
The oldest and most traditional products, bread and wine.

▶Vyta Santa Margherita Bologna. Piazza delle Medaglie d'Oro, 1b, 40121 Bologna BO, Italy / +39 051 247183

Bologna is the beginning of a new phase in VyTA Santa Margherita format: brand recognition is no longer entrusted with the application of an architectural standard declinable on various outlets, but they are the oldest and most traditional products of our table: bread and wine, the way in which they are offered to customers, the atmosphere and a design in constant change, but generated by common matrix, to represent the true essence of the brand. Each should be different and unique in its own way, with at least one new idea that sets it apart from its predecessors, Vyta Bologna is a big step in that direction, the stores should represent their brand, but they should also be a very special and unique food experience.

Vyta Santa Margherita offers the oldest and most traditional products, bread and wine, in Bologna railway station, with a C-shaped floor plan, Vyta is totally projected towards the outside through three different views: on Medaglie d' oro square, in front of the taxi station, on the historic atrium, outlet flows of underground walkways that connect to the high speed station, located at 23 metres in depth, on the so-called "Transtlantico" building dedicated to services for travelers covered by a glass and metal structure, it is exposed to the urban and metropolitan fluxes of the city, but at the same time it is separated from the never-ending come and go of the people from an invisible boundary generated by a sophisticated minimalism.

Project details

- **Design:** Collidaniel Architetto
- **Homepage:** https://www.collidaniela.com
- **Area:** 135 mq+50 mq
- **Location:** Bologna, Italy
- **Photo:** Matteo Piazza

About us

COLLIDANIELARCHITETTO is an award-winning architecture and interior design studio, founded in Rome in 2009 by Daniela Colli. The ability to combine the contemporary vision of society and its needs with the historical and cultural roots of interior design are the hallmark of projects that mix the past and the future.

His activity ranges from furniture to interior design and urban architecture, with particular attention to detail in the design and construction phases, he uses the challenging elements of each project, including the peculiarity of the site , the target and the needs of the client, as a catalysing element of his ever-evolving architectures Appreciated by international critics and considered a specialist in the hospitality sector, she works on projects ranging from bars to cafés, restaurants, hotels and spas.

▶Contacts e-mail: jobs@collidaniela.com/ Call us: +39 06 97610447

VyTA Bologna is characterized by the use of a unique colour palette: the greige, less cold than grey, less warm than beige, but highly connotative, the muted colour become a mainstay in the world of high fashion in the eighties thanks to Giorgio Armani. Greige colour is dominant in its minimalism, it amplifies the fluid outlines of the space and creates a neutral homogeneous container, dominated by the contrasts of natural 'serpeggiante' marble, glossy and matt surfaces, lights and shadows.

" Through simple products offered by Nature, such as water, wheat and fire, thanks to man' s expert hand, patience and creativity, forms and savours, aromas and flavours have been created for millennia, giving birth to bread and wine, ancient and modern nourishment for humanity." This food philosophy was the starting point that inspired the architectural concept.

Section - a

La Natura offre elementi semplici: acqua, fuoco, terra. La mano esperta, la pazienza e la creatività dell'uomo creano da millenni forme e sapori, aromi e profumi: pane e vino alimentazione dell'umanità antica e moderna

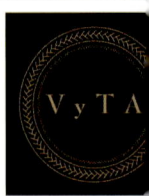

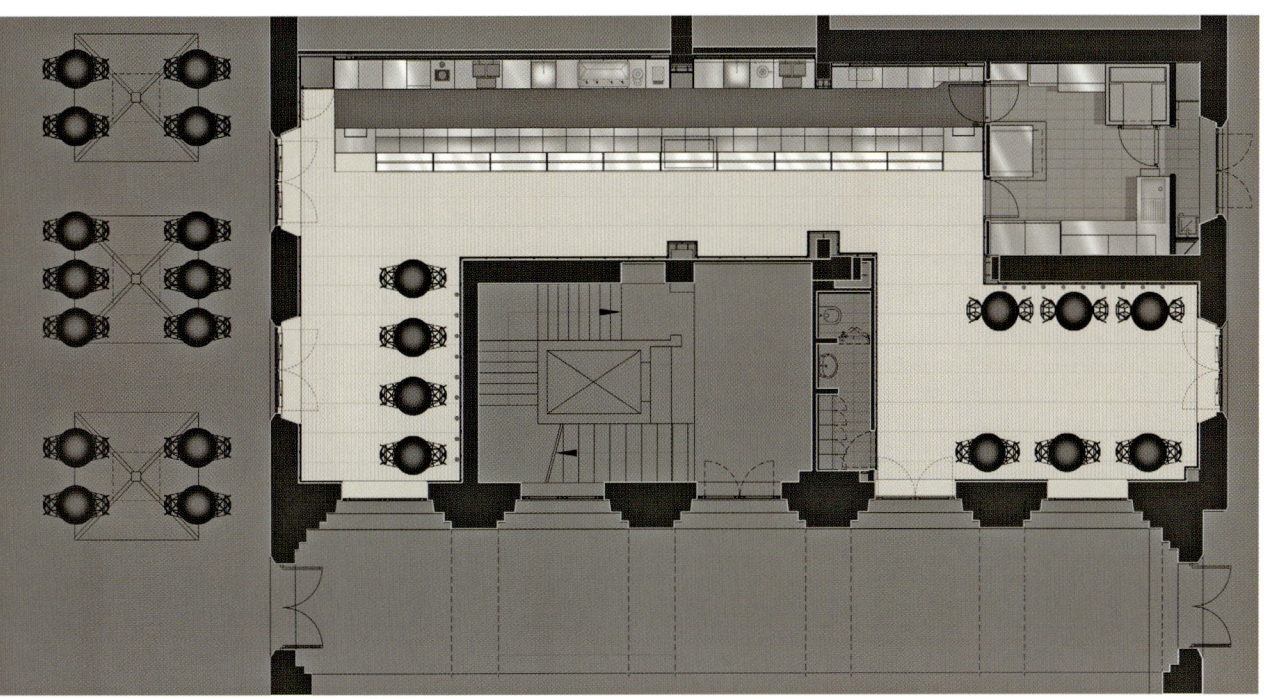

Floor Plan

111

Section - b

The long Corian elegant grey counter, with its ethereal crystal boxes, stands on the geometry of a 'serpeggiante' marble floor, the optical effect is further enhanced by mirrors which help to magnify the narrow space, the wall of the back counter, usually just a functional space, becomes a new form of entertainment through a video wall that reproduces a perpetual movie in slow motion that celebrates the ingredients of the products Vyta Santa Margherita such as water, wine, olive oil, wheat, flour, mozzarella, tomatoes and fire. The hood is one of the most significant components of the setting, due to the shape and size of its wooden lacquered greige planks that evoke the interweaving of traditional bread baskets. This volume has been brought down to a human architectural scale, so that the space has a less monumental and more intimate look.

The section of the slats, concave and convex, generates the matrix of a thin and elegant frieze: made of painted steel sheet, it becomes a micro three-dimensional sign that emphasizes the glossy Parapan background enclosing the space like a treasure chest, engraved on the mirror transforms it into a kaleidoscope of endless reflections that create an ever-changing environment.

The light system contributes to soft and intimate atmospheres: it diffusely radiates on the counter, enhances the wooden lacquered hood, it is an eye-catcher above the tables thanks to sculptural Diamond lamps by JSPR, emphasizes the shiny parapan walls with friezes metal where the bottles of wine and their history are celebrated like sculptures.

The lighting system, totally LED dimmable, merges with architecture, a sophisticated home automation system allows a resources optimization, reduces energy consumption and achieves a comfortable lighting environment, giving life to an extremely sophisticated technological universe, in harmony with seasonal conditions and the natural light on the streets outside.

The Miura tables by Plank, as well the chairs one and the stools one by Magis, with their shapes generated by the symmetrical composition of irregular triangles, can host the clients on their brief stops, while eating and drinking the most ancient products of humanity, watching the ever-moving city.

▶Holiland Patisserie: Shanghai, China / +39 051 247183

Shanghai, China
Holiland bakery

London-based practice Universal Design Studio has designed a pastel sweet bakery in Shanghai, where the fitout almost steals the show from the sweets.

For a new patisserie and bakery concept in Shanghai's Vanke Mall, London practice Universal Design Studio took inspiration from the products on display – intricately delicate cakes. The aesthetic at the newly opened Holiland is intentionally bright and youthful, while tempered by softer neutrals and natural finishes.

The material selection played a big part in the development of this project with their execution adding graphic impact and a clean backdrop to let the hand-crafted delicacies shine through. The palette is defined by a soft colour palette of oak wood, pink resin and two variations of terrazzo. Another key feature is the application of geometry. Curved and rectangular volumes pull the customer's eye in with a particular focus on the recessed display niches. The project for the retail concept encompassed designing all aspects of the interior as well as the facade. Space planning also played a major role with the layout across the 1100-square-foot space incorporating moments of retail theatre whereby customers can view the patisserie chefs as they make all of the delectable treats.

The curves are repeated throughout the space as a visual connector, but also soften what would be an otherwise a large white box. Graphic curved geometries dot the branding, while the terrazzo plinths feature chamfered edges. Making a dramatic impact is the curving, plywood-lined wall behind the counter. This envelopes the space and adds a sense of compression and release.

Project details

▶**Design:** Universal Design Studio
▶**Homepage:** https://www.collidaniela.com
▶**Area:** 103.sqm
▶**Location:** Shanghai, China
▶**Photography:** Seth Powers

About us

Universal Design Studio was founded in 2001 by Edward Barber and Jay Osgerby in Shoreditch, London. The studio is driven by a deeply held belief in the transformative power of well designed, finely crafted spaces with a process rooted in design-led strategy and research. The studio's work foregrounds the experience of the people that inhabit our spaces, with an emphasis on adaptability and rigour.
 Over the past two decades, Universal's portfolio has grown to encompass hotels and restaurants, retail environments, workspaces, residential, master planning and public realm design. Alongside this, the studio has executed culturally significant projects for galleries and cultural institutions across the globe. Clients include Ace Hotel, Fortnum & Mason, At Six Stockholm, The Office Group, Rimowa, the Victoria & Albert Museum, IBM, Google and Frieze Art Fair amongst others.

The bakery's exterior allows customers to peruse the cake selection through an external kiosk, which is offset by outdoor seating populated by greenery and planters. The blurring between interior and exterior extends to the terrazzo tiled floor finish, intended to draw the customer's eye into the space.

■

Geometry is central to the design, using curved and rectangular volumes to draw attention to each of the display niches. In creating the retail concept, we shaped all aspects of the interior, from the architectural finishes and joinery to the façade, utilising a combination of refined materials presented in a soft colour palette of oak wood, pink resin and two variations of terrazzo.

Pink me.

▶Holiland Pink Store: Jiangning District, Nanjing, Jiangsu, China

Nanjing, China
Holiland Pink Store
This store "little pink" the main style is girly and cute!

We imagined the entire space as a pink cultivation room for cultivating seeds. The seeds are cultivated in a greenhouse that can regulate the climate, bathed in bright light, and the pink is also used as a nutrient, which is transported to the roots in the form of liquid solution or gas through pipes.

We use the device as the core to construct the whole space and simulate the process of seed development and bread making through abstract expression. Translate all kinds of cultivation equipment into functional devices that can be used for display, display and sale, breaking the dimensional wall and allowing people to place themselves in the world of desserts.

The design of the theme store adds typical Chinese design aesthetics and Holiland's unique personality elements. The wall adopts a large area of pink as the background color, combined with a large area of visible glass, which not only reduces the visual fatigue caused by the large area of color blocks, but also increases the transparency of the space, and visually gives people Warm, girly feel.

Project details

▶**Design:** Das Design

▶**Homepage:** https://www.dasdesign.cn

▶**Area:** 235 m2

▶**Location:** Jiangning District, Nanjing, Jiangsu, China

▶**Photography:** Seth Powers

About us

Since its establishment, DAS has accumulated various types of projects covering 120 cities, continuously enabling the team to maintain a leading and international vision and design creativity. At the same time, it has also been recognized by many customers

Section-a

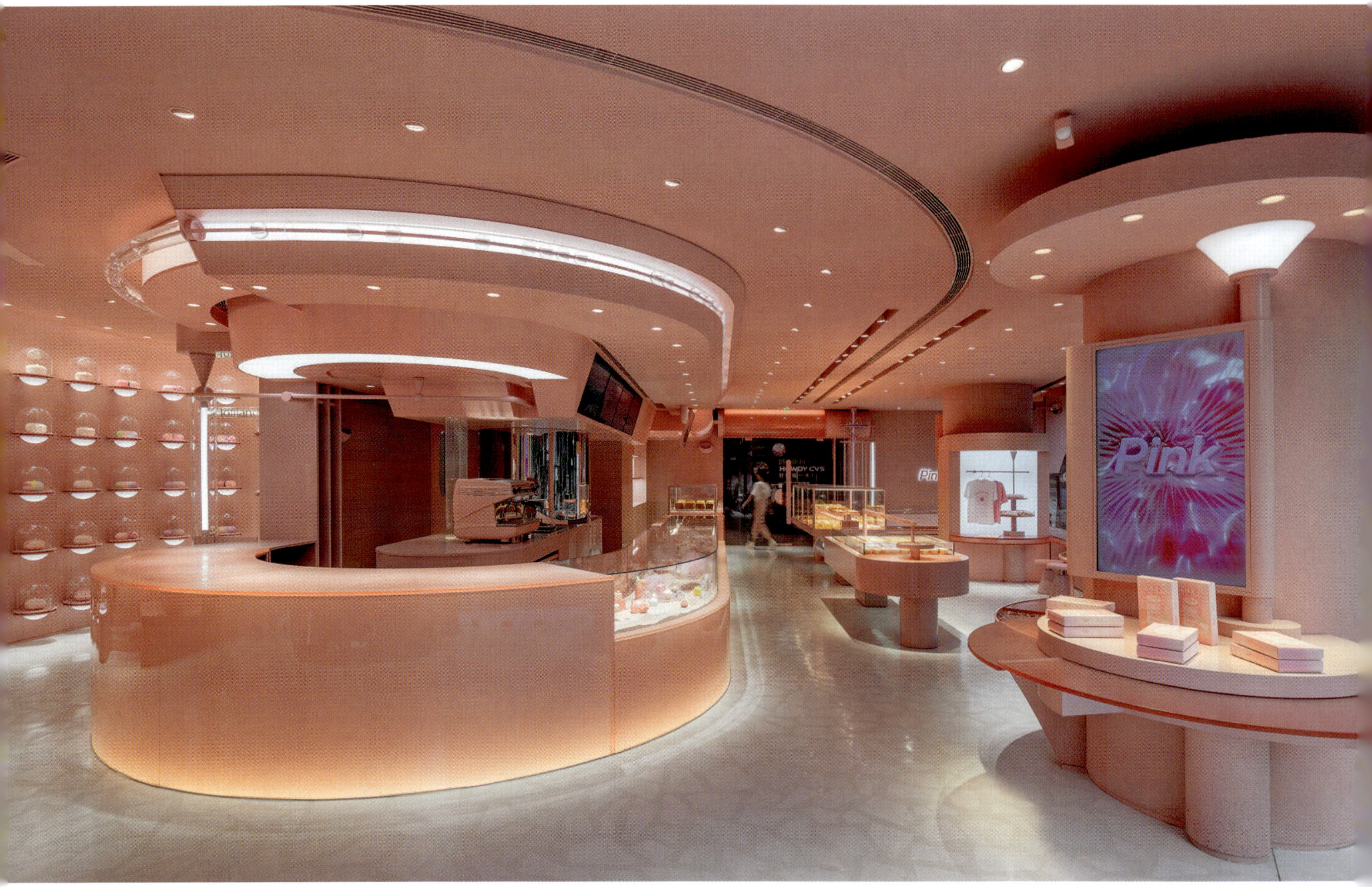
Pink

Holliand Pink
Pink me
Holliand Pink

Pink me

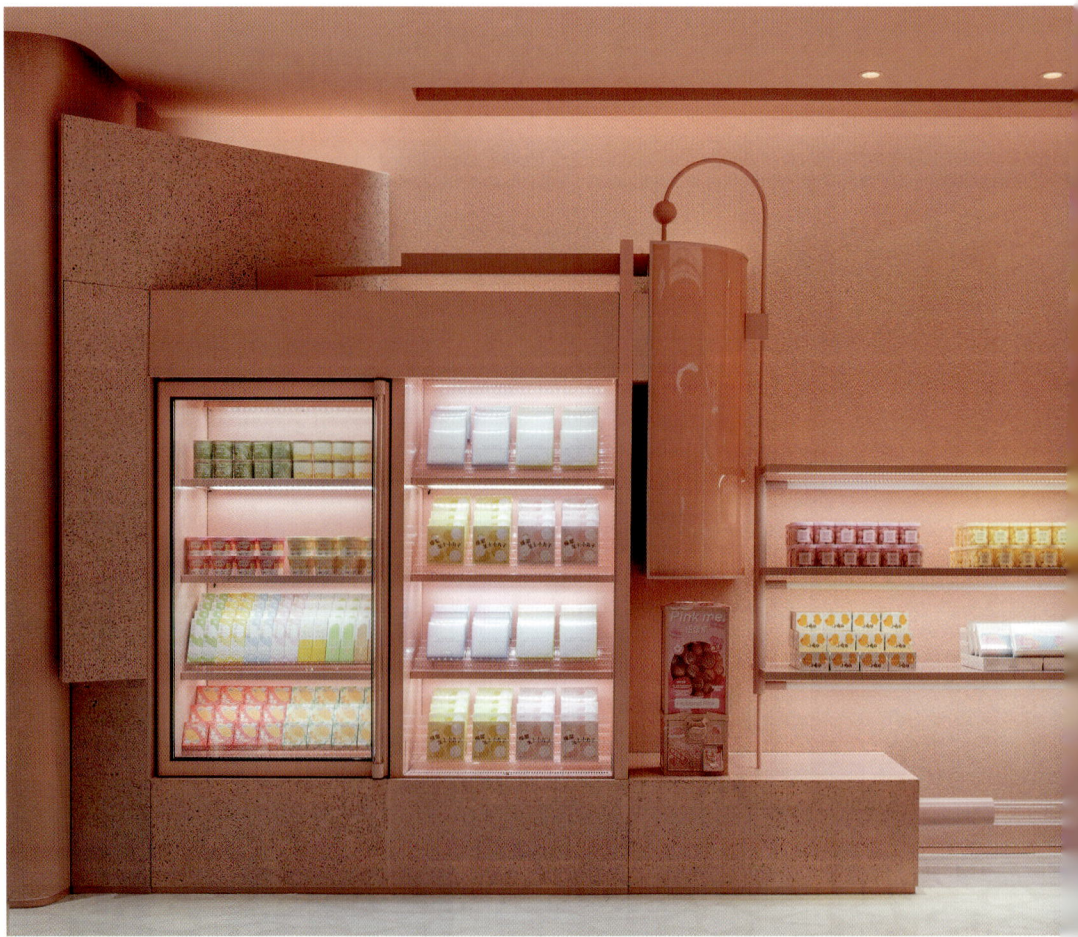

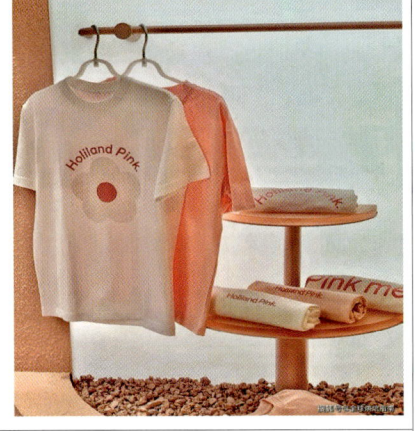

Section-b

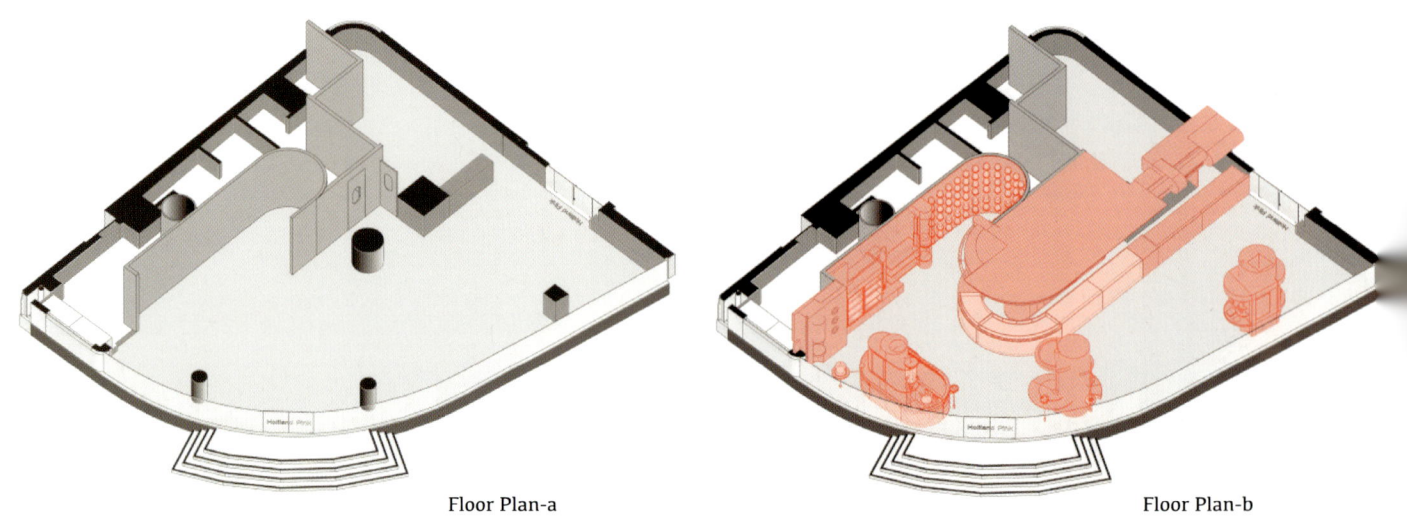

Floor Plan-a   Floor Plan-b

130

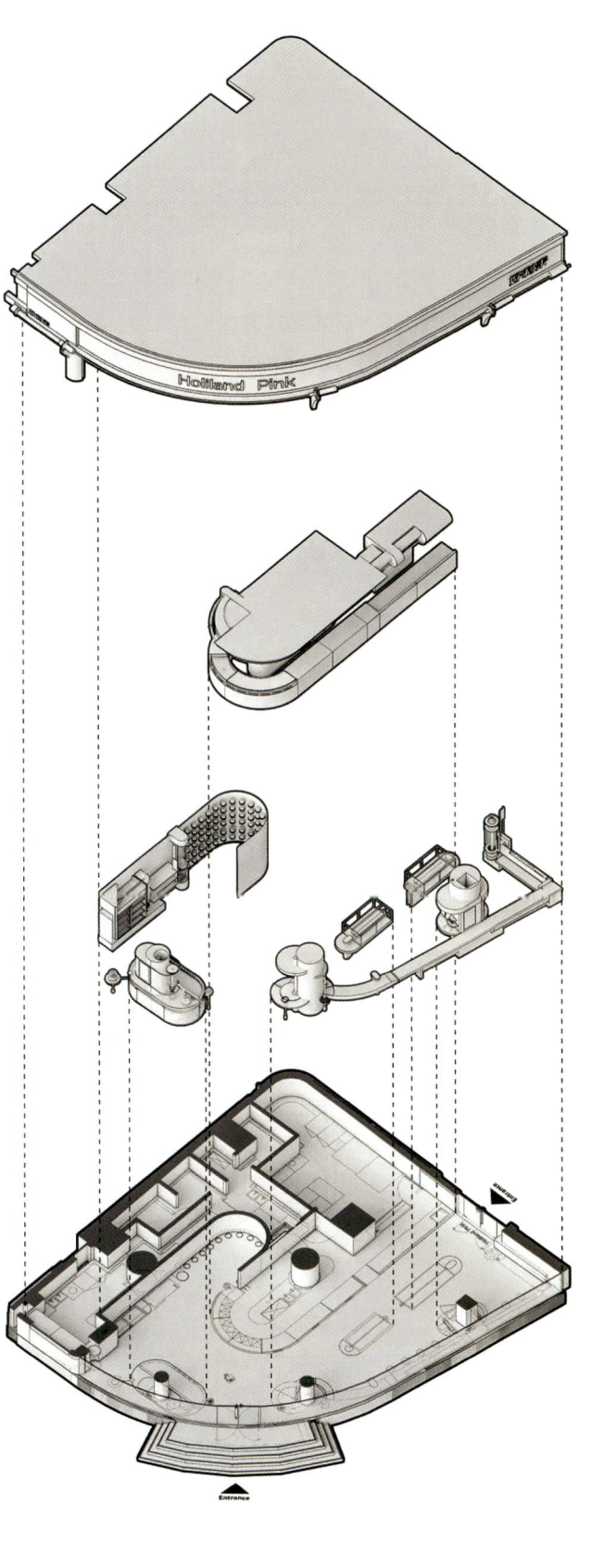

AXONOMETRIC 0 1 5m

133

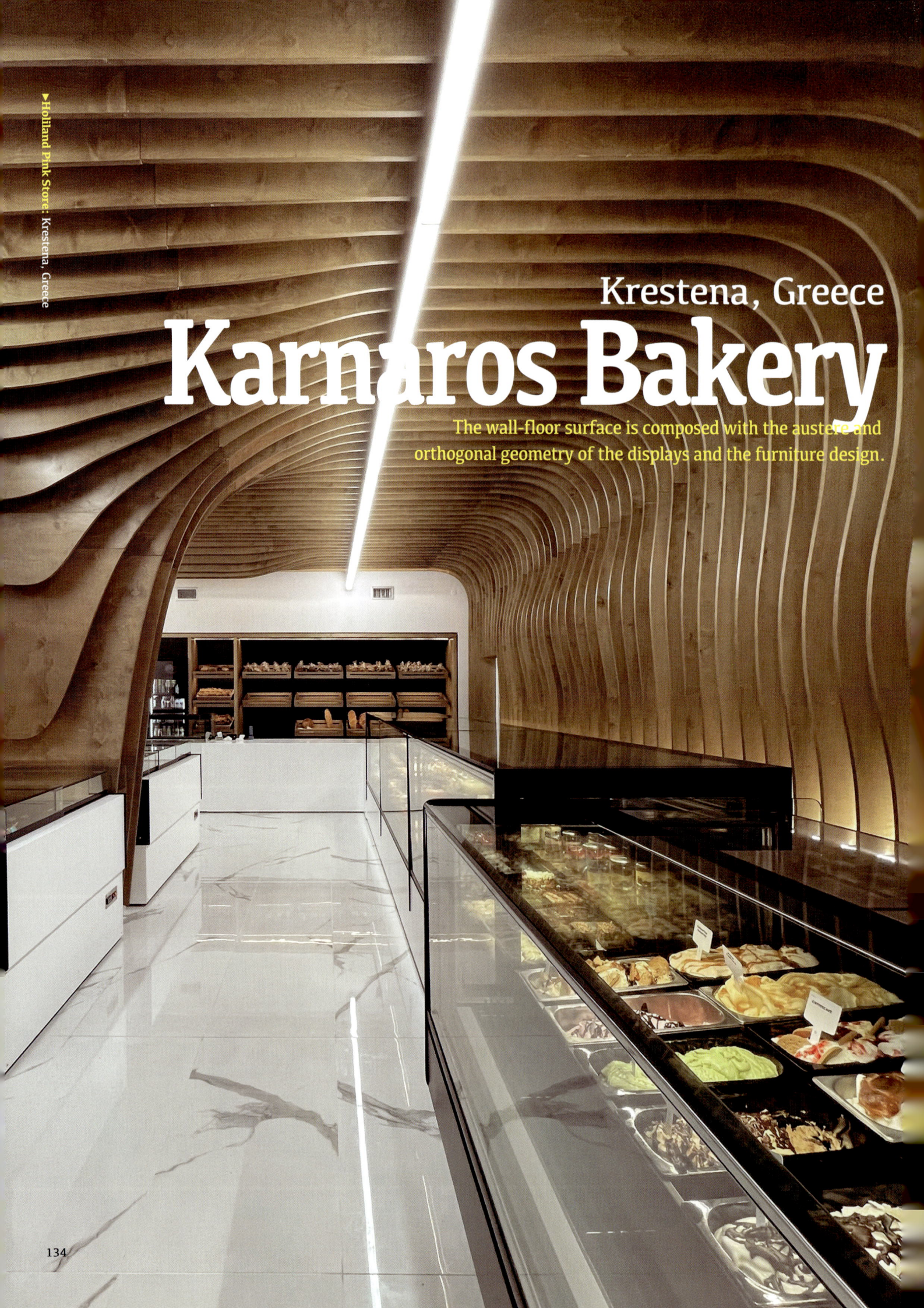

▶ Holliand Pink Store: Krestena, Greece

Krestena, Greece
Karnaros Bakery

The wall-floor surface is composed with the austere and orthogonal geometry of the displays and the furniture design.

The project concerns the redesign of a bakery & patisserie storefront with the aim to introduce a space with unique identity. The principal design axis was the synthesis of opposing pairs on multiple levels. This approach stems on the one hand from the desire to create a space that is contemporary but also familiar and intimate, and on the other hand from the combined identity of the bakery and the patisserie. Guided by this approach, there are four opposing synthetical pairs: Pair One is the curved and orthogonal: The inner curvilinear "skin" which is applied on the wall-floor surface is composed with the austere and orthogonal geometry of the displays and the furniture design.

Pair Two is the intricate and the subtle: The natural and rich texture of the birch is combined with the pure and subtle texture of the corian and the marble. The former abstractly relates to the texture of the bread and the latter to the one of the cream.

Pair Three is the saturated and the pure: The color saturation of the upper part with the color neutrality of the lower part of the space. The upper to intensify the identity and the lower to provide a monochromatic canvas for the display of the colorful products.

Pair Four is the function and the identity: A special element that was taken into consideration was the existence of a pillar in the middle of the available space, which was rendered as a functional flow problem. Instead of hiding this "necessary problem", the proposal intensifies its presence, setting it as the core of the geometric deformation. Thus it converts this element from a functional problem into a singular identity element of the space.

Project details

▶**Design:** ARCHE Architecture & Design Lab ☒Design
▶**Homepage:** http://archelab.gr/the-studio
▶**Area:** 219 m2
▶**Location:** Jiangning District, Nanjing, Jiangsu, China
▶**Photography:** Christos Dionisopoulos

About us

ARCHE Architecture & Design Laboratory is a creative studio of Architecture and Design.

Mina Sarantopoulou is an interior and product designer with studies at the internationally renowned Central Saint Martins, University of The Arts, London, from which she graduated with distinction. Her work has received important distinctions such as LogoPlastic Award, StarPack National Awards, Vetrerie Bruni Progetto Millenio European Competition. He has worked in major interior design, decoration and object design offices including Hector Serrano and Afroditi Krassa in London.

Vasilis Stroumpakos is an architect and educator, with studies at the Department of Architecture, Aristotle University of Thessaloniki and M.Arch from the Architectural Association in London. He was professor at Architectural Association School of Architecture at undergraduate and postgraduate level, and currently he is Assistant Professor at the Department of Architecture, University of Patras, Greece. His work has been featured in international exhibitions such as London Biennale, Possible Futures, Arco the Greek Suspense Madrid, Milan Bienale Beyond Media, National Museum of Contemporary Art Athens and has been published in major media such as AD, Blueprint, Spazio Architectura, World Architecture Review. He has received international awards such as Feidad, Plecnik Institution, European Design Award, AVA Digital Awards, Design Licks.

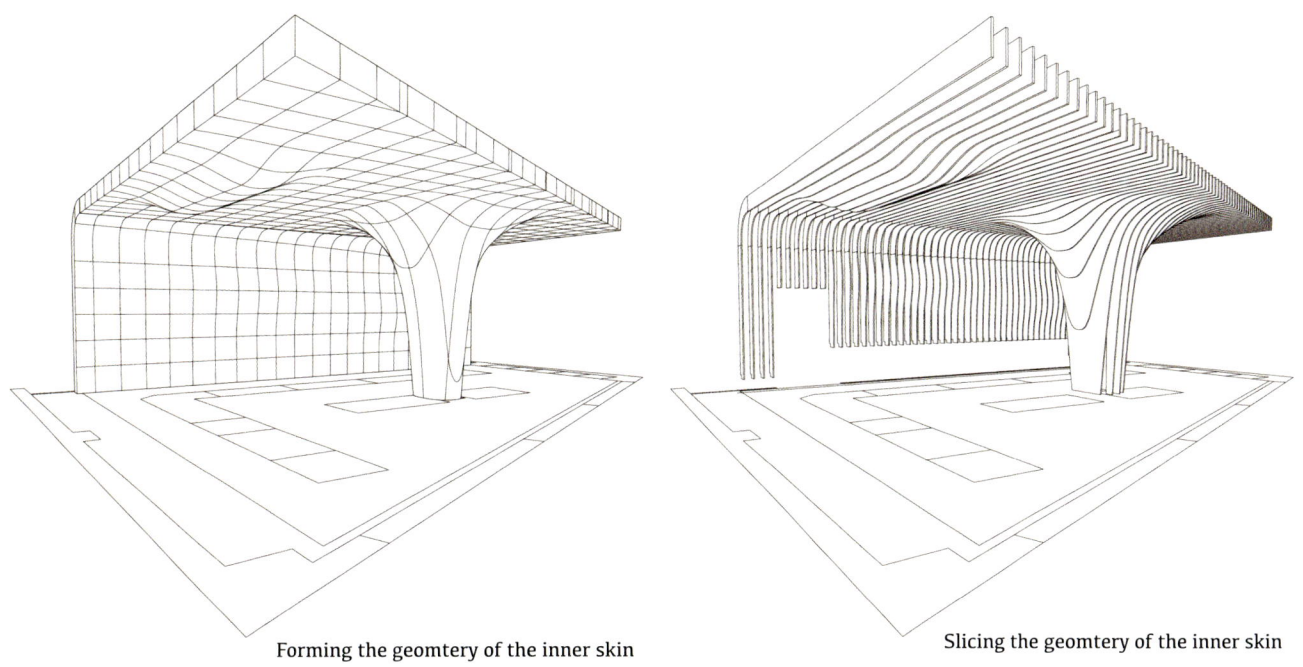

Forming the geomtery of the inner skin

Slicing the geomtery of the inner skin

137

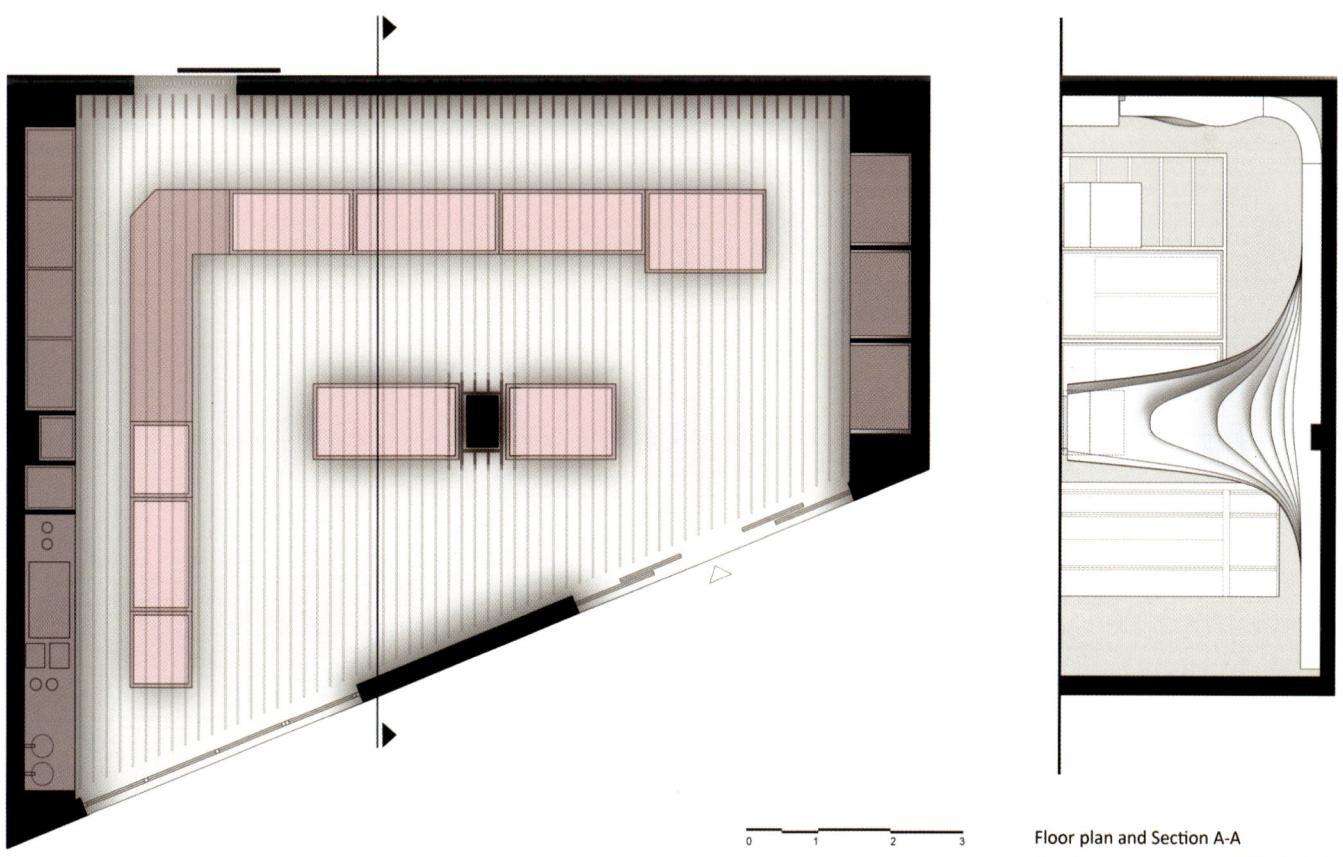

Floor plan and Section A-A

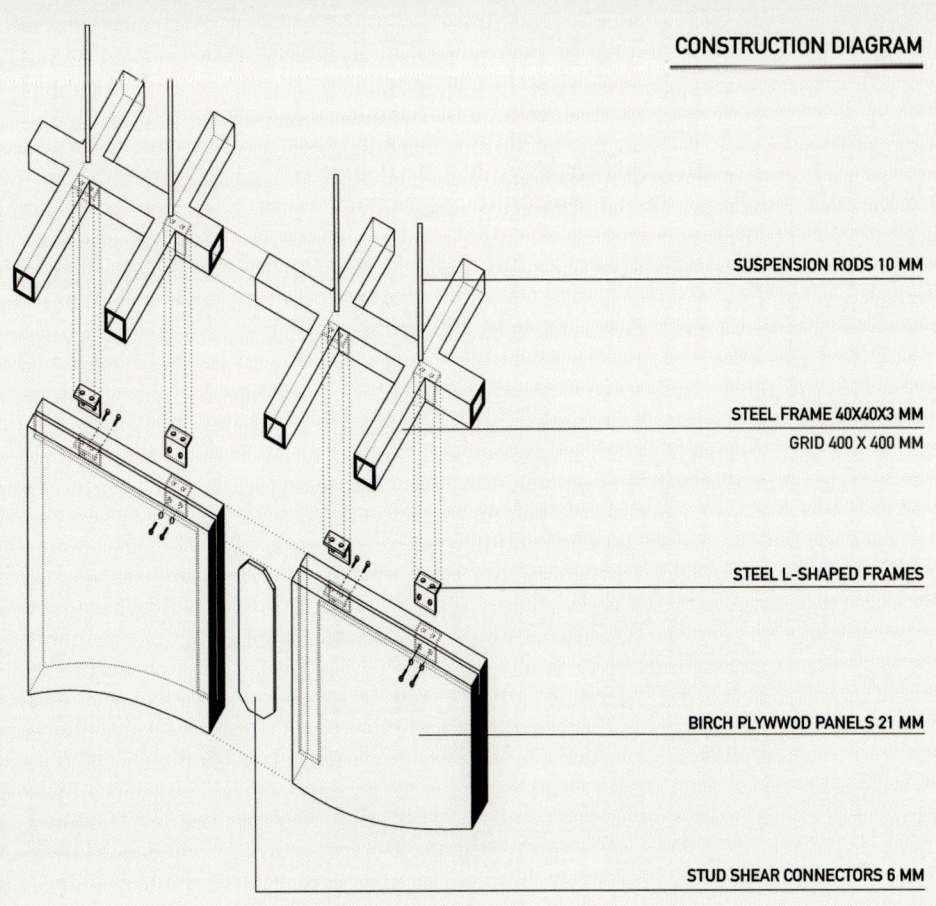

CONSTRUCTION DIAGRAM

SUSPENSION RODS 10 MM

STEEL FRAME 40X40X3 MM
GRID 400 X 400 MM

STEEL L-SHAPED FRAMES

BIRCH PLYWWOD PANELS 21 MM

STUD SHEAR CONNECTORS 6 MM

Dubai, United Arab Emirates
Dessert Moishi

Carole Mouawad is the woman behind Moishi, brand dedicated to a Japanese dessert – homegrown in the UAE

Dessert MOISHI is a luxurious Japanese confectionery brand where authenticity and craftsmanship are merged with modernism to create a refreshing and capricious taste experience. The company produces mochi ice cream and other Japanese confectionery in the United Arab Emirates. Raw materials are partially imported from Japan and met with locally-crafted ingredients to satisfy even the most demanding taste buds. The client wants the concept to be inspired by the Japanese Traditions and should reflect the variety of its products in the design – colorful, authentic, and playful.

The flooring is inspired by the Japanese Zen Garden. Zen creates a miniature stylized landscape through carefully composed arrangements of rocks, water features, moss, pruned trees and bushes, and uses gravel or sand that is raked to represent ripples in the water. Thus the floor pattern was created to replicate the same ambiance. Terrazzo floor represents the sand and gravel while the painted pattern gives the lines of Zen garden. The tables at M' OISHI is designed by 4SPACE with an inspiration from the wooden mallet used in making traditional rice cakes.

The Back wall of MOISHI is inspired by Shoji – a door, window, or room divider consisting of translucent paper over a frame of wood. The wall art murals are inspired from the Kimono floral patterns that signify tranquil ambiance and gives depth to the interior. M' OISHI is using Bonsai Vegetation in the shop as it is considered as traditional Japanese trees. Bonsai "tray planting" is a Japanese art form using cultivation techniques to produce small trees in containers that mimic the shape and size of full-size trees. The ceiling light feature of MOISHI was designed as inspired by Japanese chopsticks made of bamboo wood.

Project details

- ▶ Design: 4SPACE
- ▶ Homepage: https://4space.ae
- ▶ Area: 102 m2
- ▶ Location: Dubai, United Arab Emirates
- ▶ Photography: 4SPACE

About us

Originally established in Damascus in 2001, founders, Firas Alsahin and Amjad Hourieh, moved their practice to Dubai to be at the centre of this vibrant market.

The emirate's booming growth in the commercial sector was an impetus for the firm to explore all the opportunities in the design industry.

Overcoming an uphill battle, 4Space Design has gone on to create noteworthy projects in the UAE. Eschewing quantity for quality, profile of the project and relationship with clients, the studio credit its people' s distinct ideas strategic business development.

Were you familiar with Japanese desserts before coming up with your concept?
I have always been passionate about food, but I was more familiar with Japanese cuisine than desserts specifically.

What made you want to develop your own brand of mochi ice-cream?
I fell in love with the idea of mochi ice cream the very first time I tried it. I would say it was my love for ice cream and Japan's unique culture as well as creating beautiful and playful desserts that drove me to launch M'OISHÎ. I wanted to add some beauty to this world with tasty desserts!

Carole Mouawad is the woman behind Moishi, brand dedicated to a Japanese dessert – home-grown in the UAE. If you're unfamiliar with desserts from Japan, Mouawad's concept is derived from mochi, which is a traditional Japanese rice sticky dough, and "mochi ice-cream" is when ice-cream is enveloped within a layer of mochi to make a delicious sweet treat.

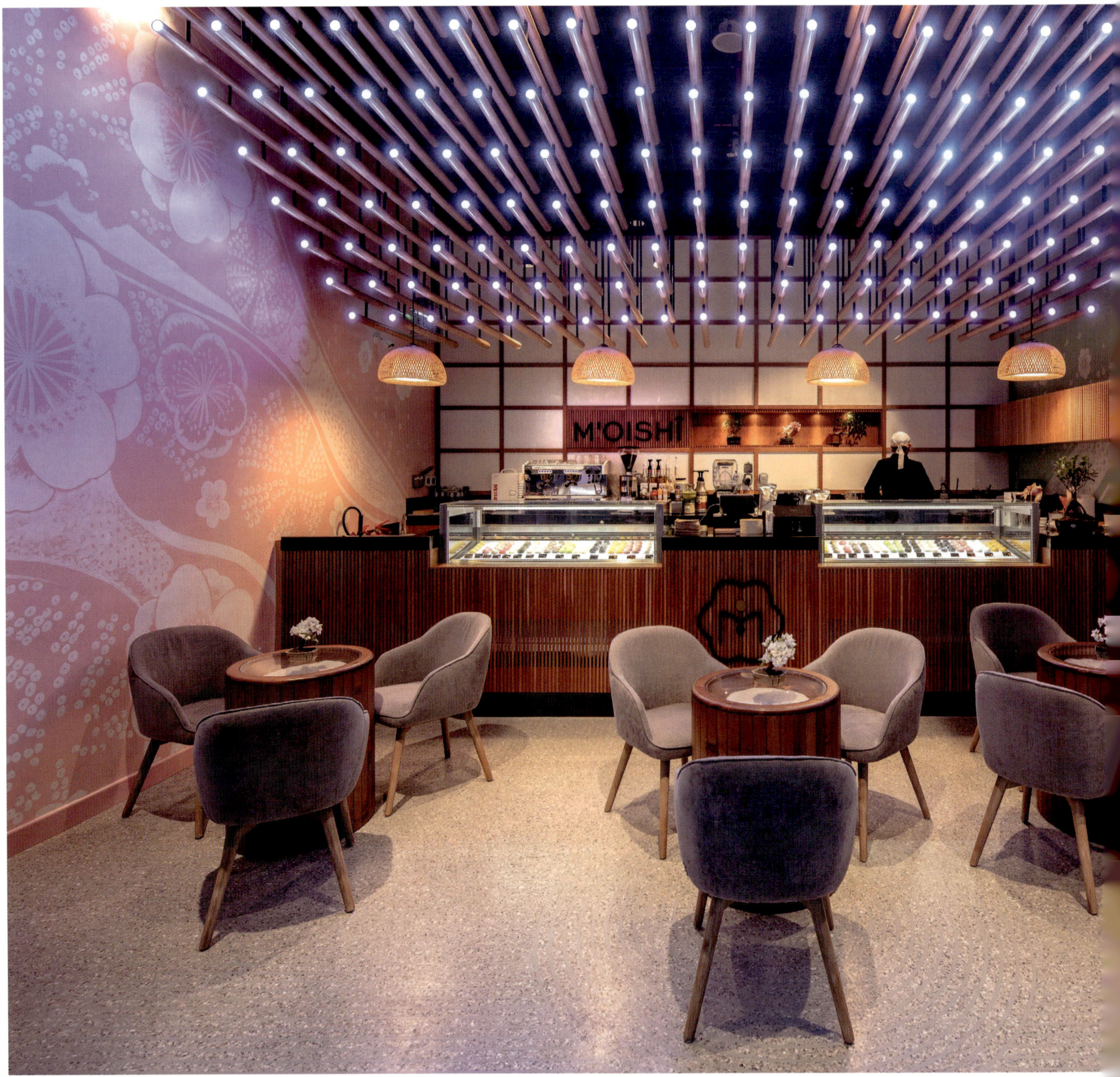
M'OISHI

M'OISHI
M'OISHI

M'OISHI

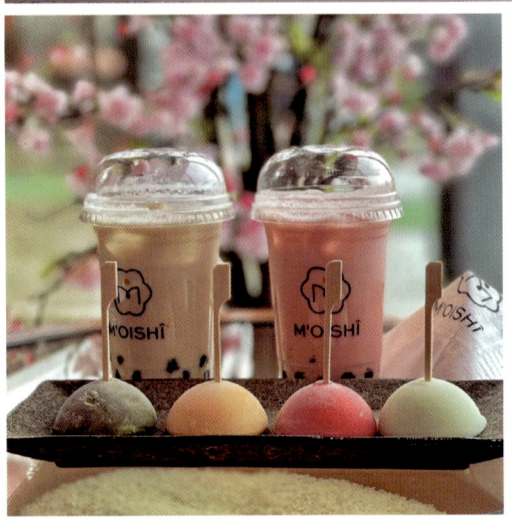

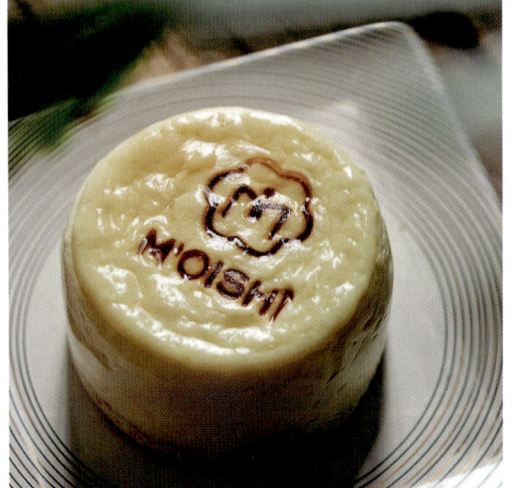

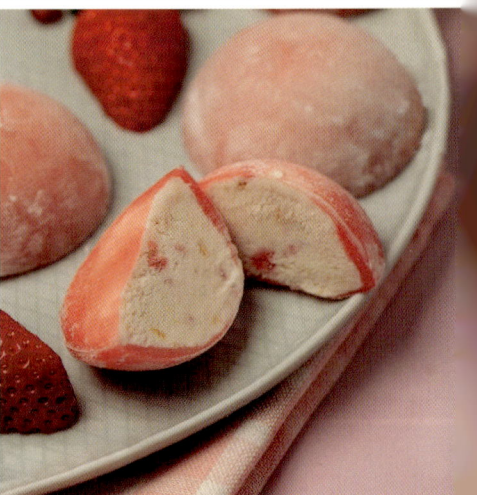

To integrate the Mochi ice cream and its flavors in the design, 4SPACE thought of producing a bamboo led lighting feature with changing colors so it can create a visual excitement and public engagement. The different colors represent all the flavors of the Japanese ice cream and the rounded lights signify the shape of Mochi. The bamboo rods were created with a dimming light technology that changes the café mood every minute. It delivers an exceptional piece and creates an ideal moment.

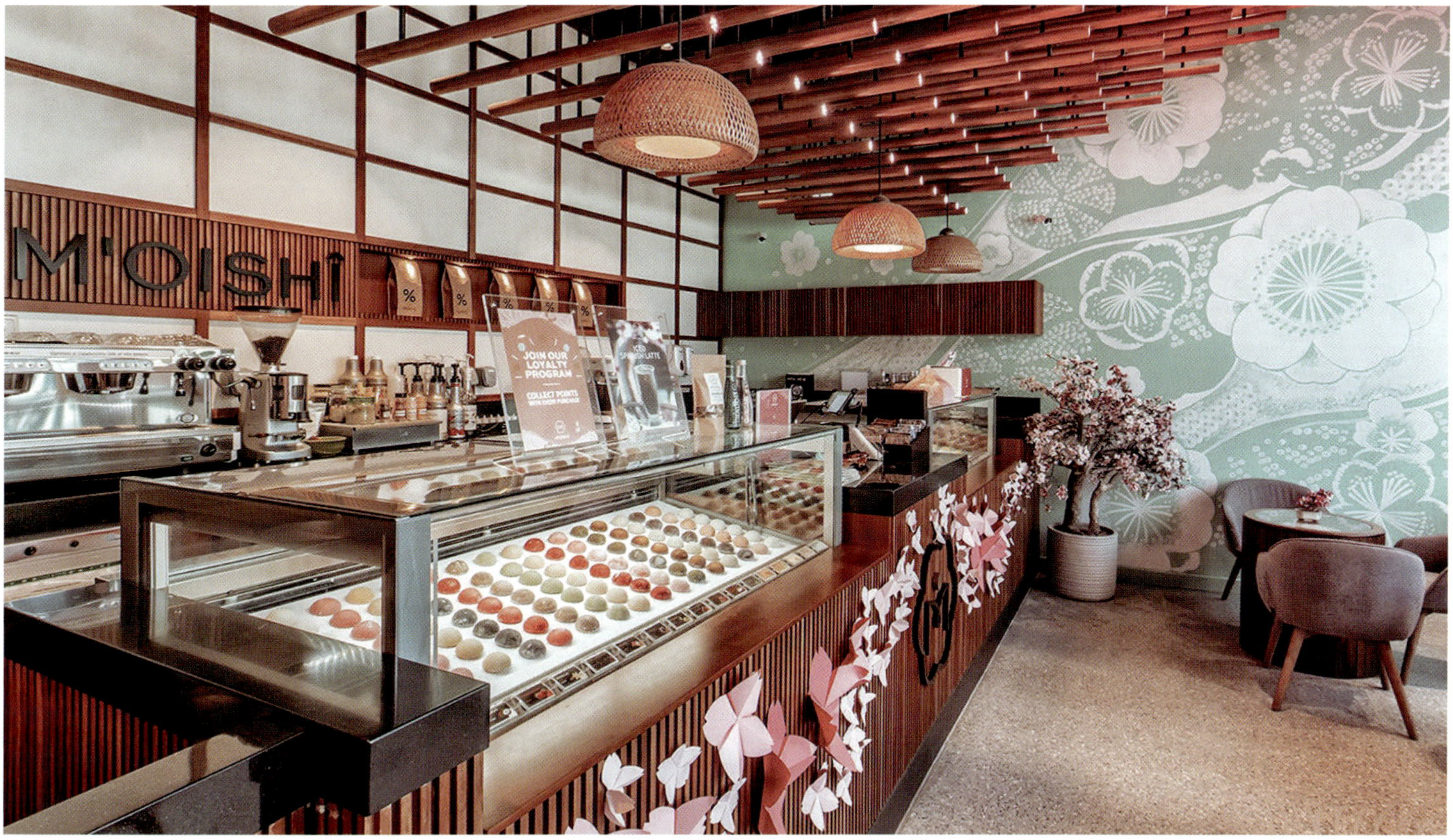
M'OISHI
JOIN OUR LOYALTY PROGRAM

M'OISHI

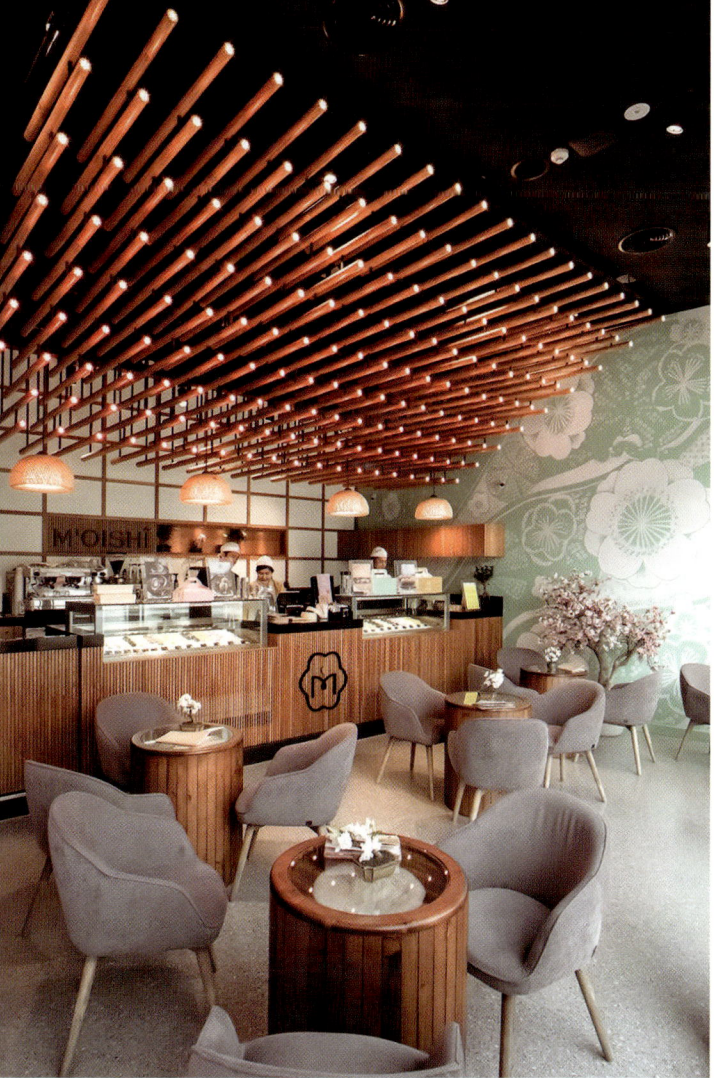

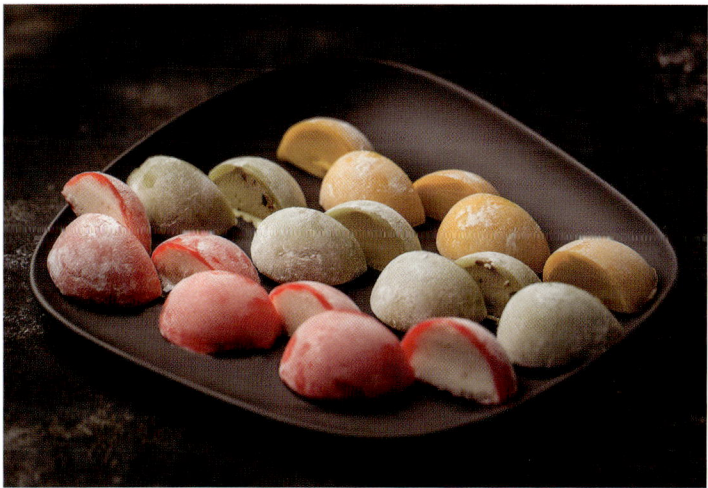

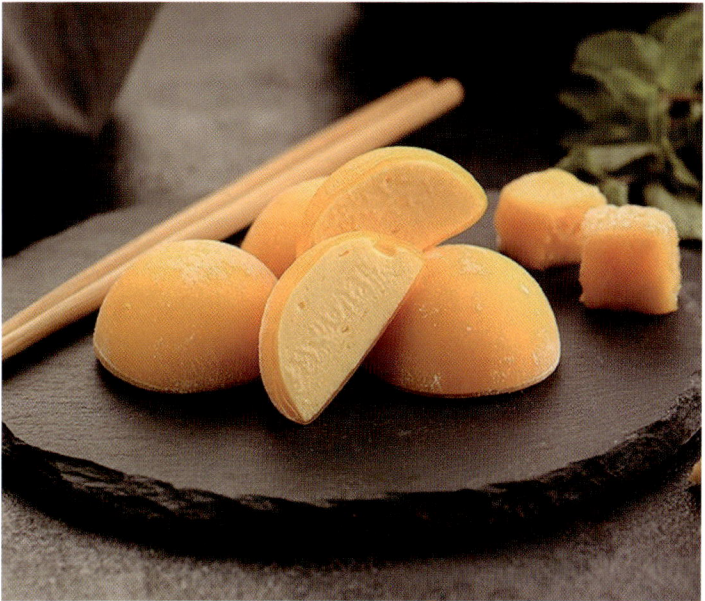

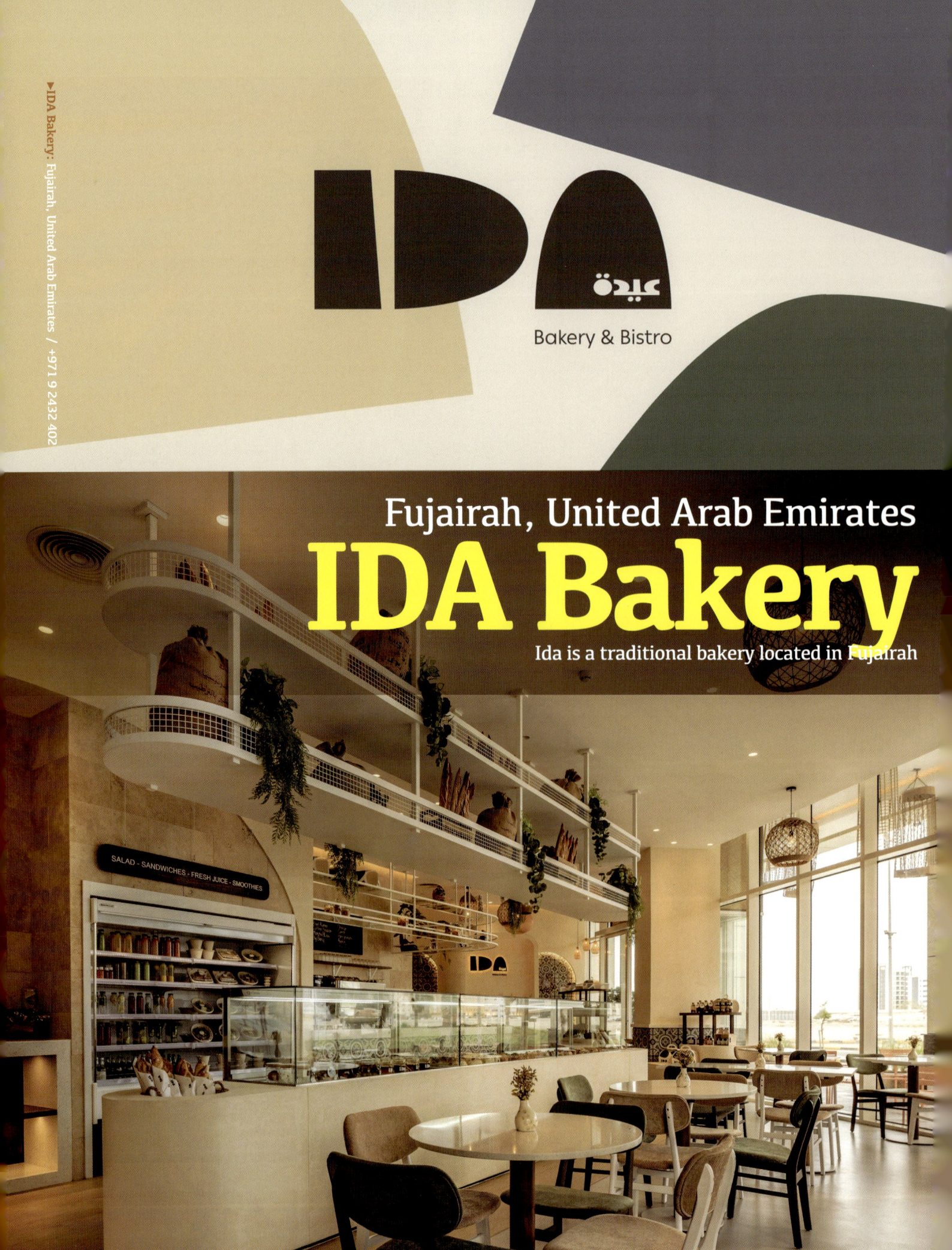

IDA Bakery

Fujairah, United Arab Emirates

Ida is a traditional bakery located in Fujairah

We Are Crafters Of The Fresh, Giving Wholesome Ingredients A Delightful Twist. Craving The New, We Never Stop Experimenting, From Baking Our Customers Original Varieties And Flavours Of Bread To Exploring New Dishes. Every Day, We're Busy Refining The Complete Dining Experience. We Can't Help But Share Our Love For Artisanal Food. Our Pursuit Of The Perfect Wafts Through The Air From Our Bread Ovens, It Surprises Through Flavourful Combinations Of The Freshest Ingredients, And Can Also Be Seen In The Pride We Take In Making Our Home Our Customer's Home. We Live For Natural Simplicity. There's An Honesty To How You'll Find Us, Always Open And Welcoming. It Influences All Our Decision Making, Right Down To How We Carefully Consider Where We Source Our Ingredients.

Ida is a traditional bakery located in Fujairah that creates a variety of gourmet delicacies using organic ingredients. The idea was inspired by raw wheat and flour. The bakery is designed with a homey atmosphere in mind. The goal is to make customers feel more at ease in a neighborhood corner shop.

Project details

▶ **Design:** 4SPACE
▶ **Homepage:** https://4space.ae
▶ **Area:** 102 m2
▶ **Location:** Fujairah, United Arab Emirates
▶ **Photography:** Anas Rifai

About us

Originally established in Damascus in 2001, founders, Firas Alsahin and Amjad Hourieh, moved their practice to Dubai to be at the centre of this vibrant market.

The emirate's booming growth in the commercial sector was an impetus for the firm to explore all the opportunities in the design industry.

Overcoming an uphill battle, 4Space Design has gone on to create noteworthy projects in the UAE. Eschewing quantity for quality, profile of the project and relationship with clients, the studio credit its people's distinct ideas strategic business development.

Our passion for baking using the finest ingredients is what fuels our journey at Ida.

We are creators of the exquisite, daily. We love nothing more than bringing the best quality baked goods and wholesome food to people in our communities. Wherever we are, we give our customers a taste of home away from home. Every day sensational.

From Baking Our Customers Original Varieties And Flavours Of Bread To Exploring New Dishes.

▶Cloud & Co.-Doha: street 930, Doha, Doha, Qatar / +974 3003 0433

Cloud & Co.-Doha
Doha, Qatar

Cloud & Co. is a traditional gelato store located in Qatar.

As intensely sweet on the eyes as their produce tastes, studio futura has stirred quite a treat with cloud and co-, a dreamy gelato shop in msheireb downtown doha, qatar. the designers created a pastel combo of eye-popping colors splashes with abstract geometries, and an almost surreal universe unravels. moreover, the visual continuity, the stunning details like the metallic decorations and mirrors, along with the theatrical curtains and seats, all come together to give gelato lovers an unforgettable experience.

'We believe that the only way of being relevant nowadays is to create authentic brands that stand out locally and globally connecting and developing lasting impact with people', as the designers explain. Studio futura was founded in 2008, in mexico city, with the aim to provoke traditional design and to push design boundaries. as such, the infamous illusionary drawings of illustrator maurits cornelis escher inspire the branding and interiors of the store. the designers take the distorted realities of the artist and turns the impossible into a tangible dream. with a combination of level changes, bold symmetries, monochromatic surfaces and carefully placed lights, a seductive new gelato culture emerges.

The space is divided in two sections. one is primarily designed in shades of cotton candy pink, incorporated with interactive elements that transforms the customer's area into a playful and imaginative atmosphere. the other, in blue and green combinations, gives off more mysterious vibes. the two areas are connected by a serving counter, featuring the gelato specials of the day. for extra comfort, seating steps and stools complete the room.

Project details

- **Design:** Studio futura
- **Homepage:** http://byfutura.com
- **Area:** 160 m2
- **Location:** street 930, Doha, Qatar
- **Photography:** Studio futura, Rodrigo Chapa

About us

In a world full of dull brands that look alike, the ones that stand out and create powerful relationships are the ones that succeed and last. Futura is a creative studio based in Mexico City, founded in 2008, with the goal of transforming traditional design forms and trying to change the way design is developed and consumed globally.

Our work is the result of constant experimentation, we believe in creativity above all else and how it can be translated not only in images, but also in objects and spaces. The essence of Futura is in making perfection meet chaos. We create and revamp brands that stand out to create meaningful relationships with people, empower businesses and resonate around the world. We believe that the only way of being relevant nowadays is to create authentic brands that stand out locally and globally connecting and developing lasting impact with people.

▶Contacts e-mail: new@byfutura.com / Call us: + 52 (55) 8661 9898

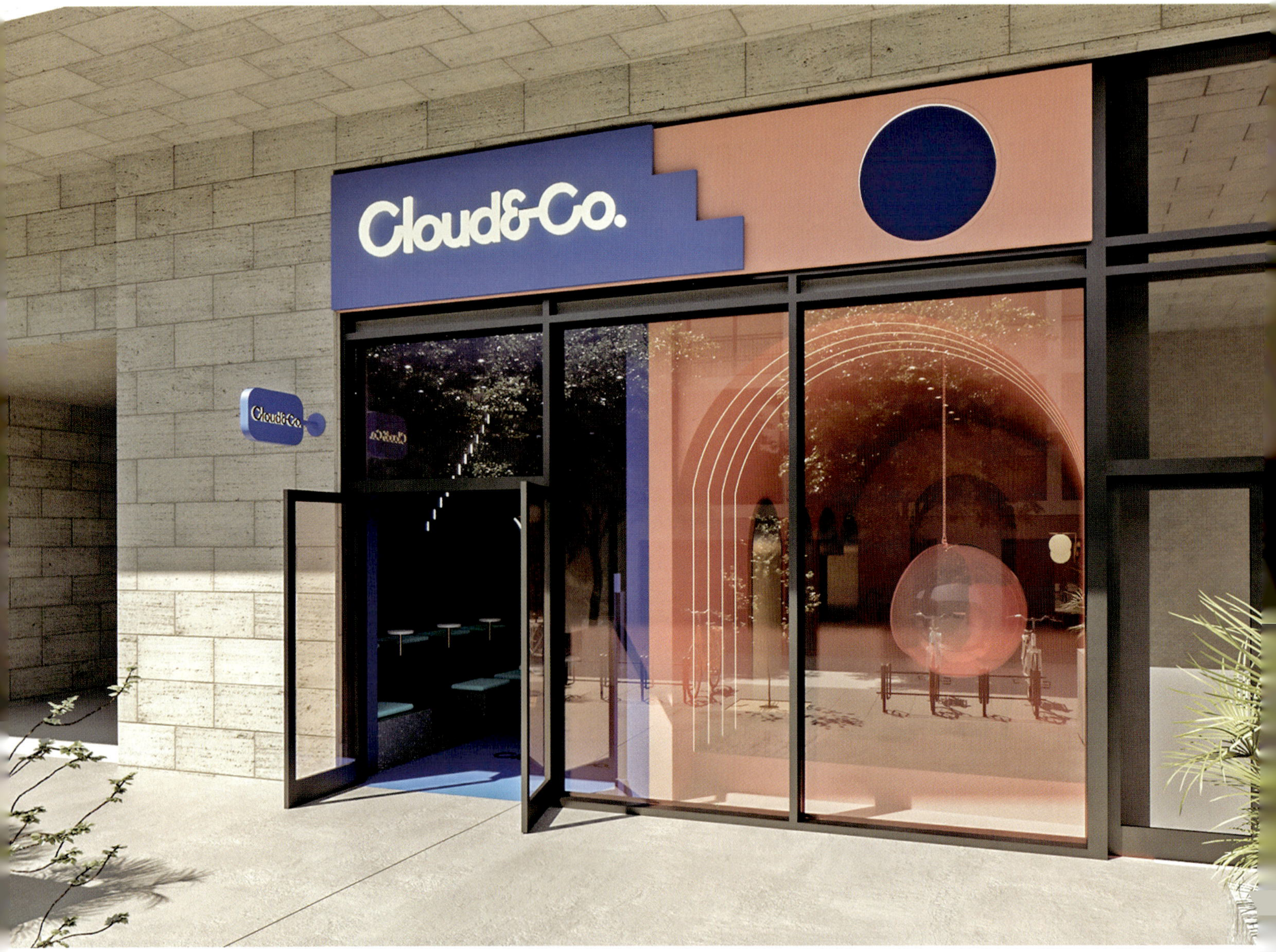
Cloud&Co.

Cloud&Co.

Cloud&Co.

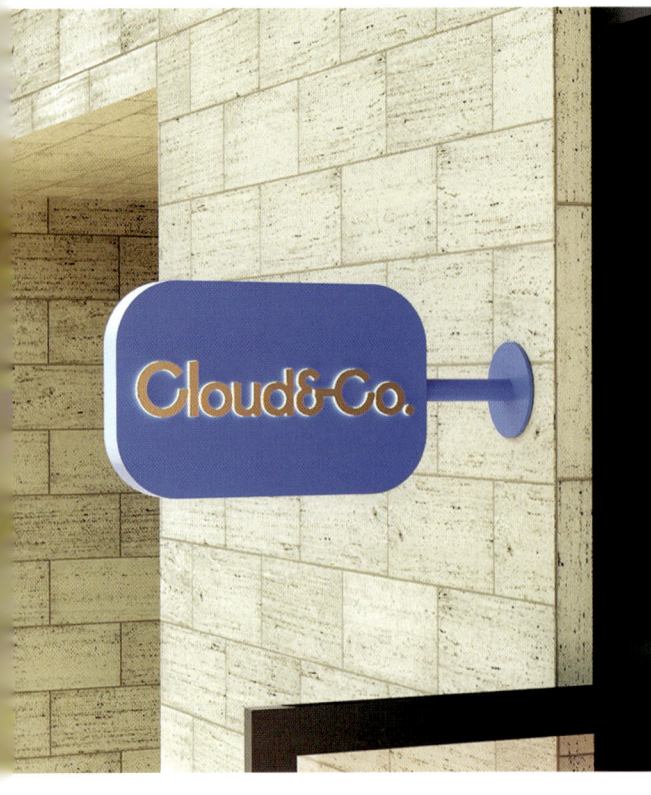

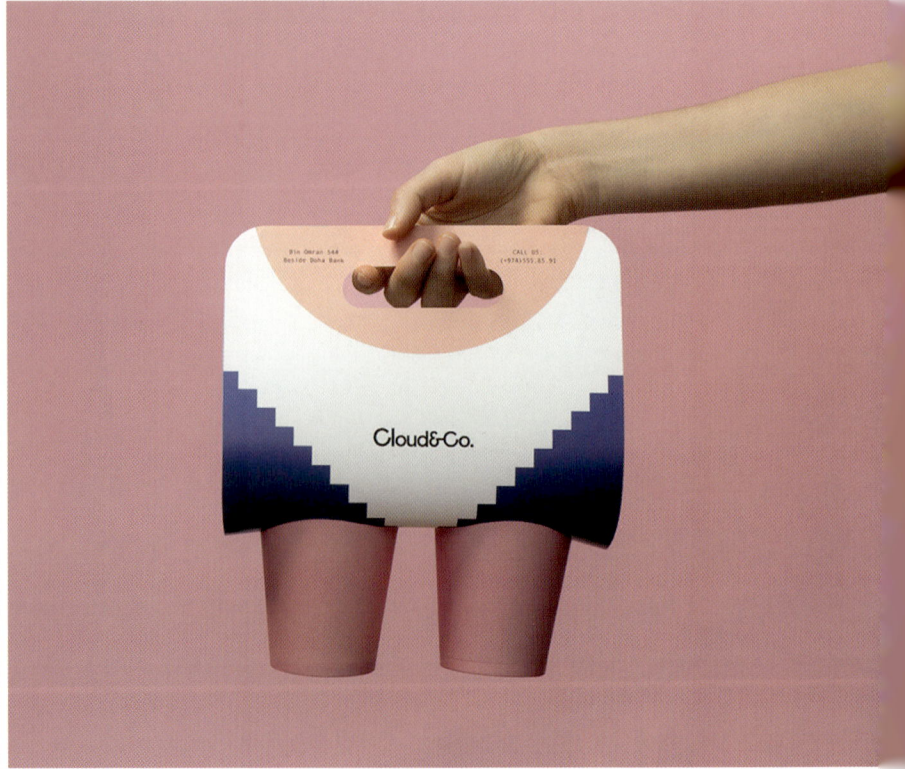

Taking as inspiration the impossible scenes of Escher, we create illustrations where everything is possible, where your dreams come true, where things without sense are the norm. A fantastic world of pastel skies and cotton candy clouds. We also use geometric figures and eye catching colors, that combined with a minimalist but fun logo, that complements the intricate illustrations.

The interior design followed the inspiration of Escher' surreal scenes and refers a dreamlike universe to reaffirm the new gelato culture in a fantastic atmosphere of pastel skies and cotton candy clouds. The customers' area is divided into two parts: one completely pink in which an imaginative and playful world is created incorporating elements that encourage the interaction. On the other side, a totally blue environment creates a mysterious and dreamy scene. Both parts are connected by a counter that manages to unite both areas into a daydream world.

The worktop, front paneling and counter sides are made of Corian Whitecap in a matt finish. The operator side was epoxy powder coated in traffic white RAL 9016 color in a matt finish.

Cloud & Co. is a restaurant located in Qatar, serving a selection of Gelato, Cafe, Desse

at delivers across Musheireb.

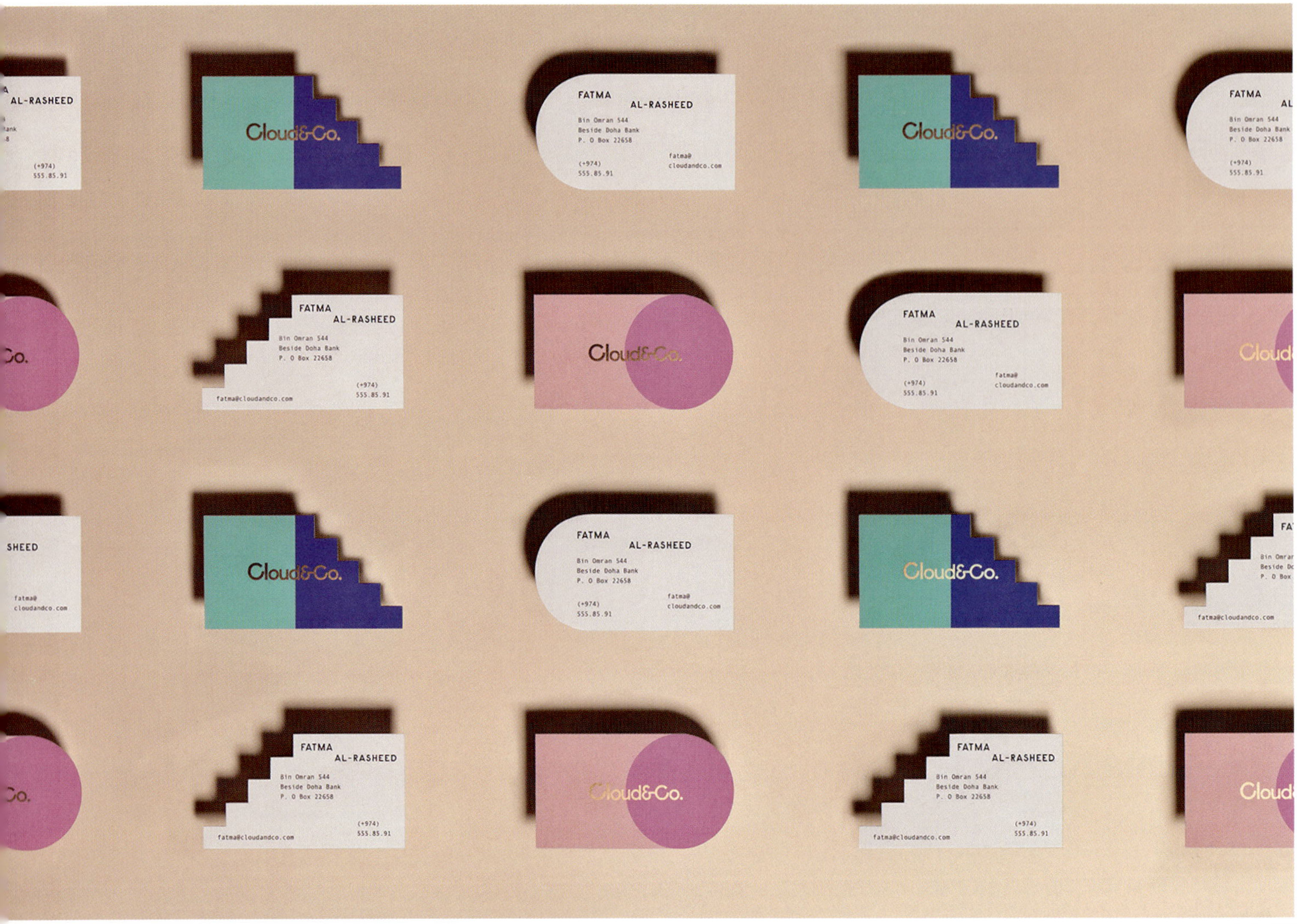
Cloud&Co.
FATMA
AL-RASHEED
Bin Omran 544
Beside Doha Bank
P. O Box 22658
(+974)
555.85.91
fatma@
cloudandco.com
fatma@cloudandco.com

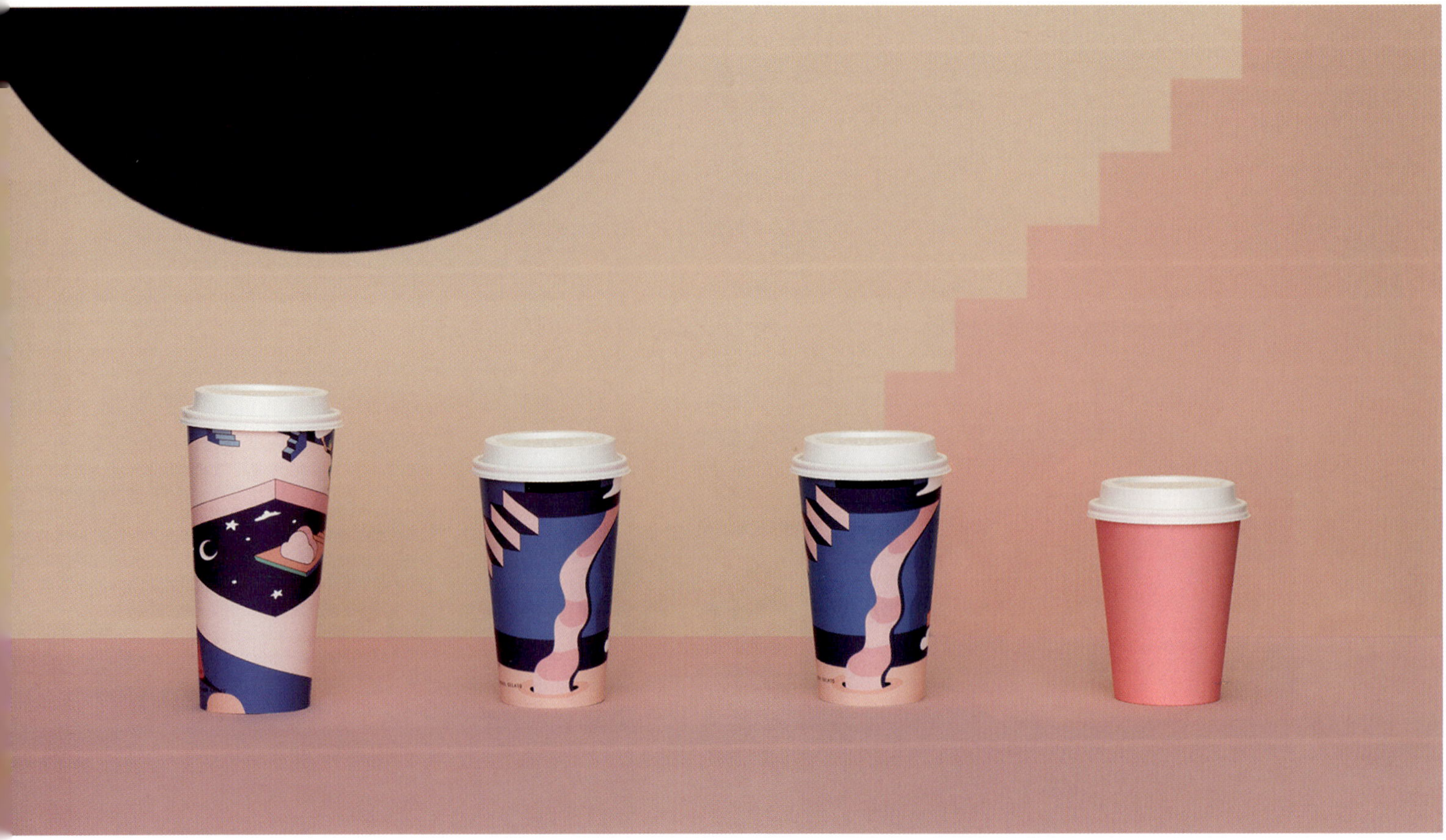

Gelato & Coffee
$ 4.00
PICCOLO
4 oz. -Unlimited Flavors
$ 4.75
MEDIO
6 oz. -Unlimited Flavors
$ 5.50
GRANDE
8 oz. -Unlimited Flavors
$ 8.50
PINTA
16 oz. -Unlimited Flavors
$ 16.00
QUARTO
32 oz. -Unlimited Flavors
ESPRESSO
Single/Double
$3.50 / $4.25
CAPPUCCINO
Regolare/Double
$3.50 / $4.25
LATTE
Regolare/Double
$3.50 / $4.25
AMERICANO
Regolare/Double
$3.50 / $4.25
BREVE
Regolare/Double
$3.50 / $4.25
CLOUD & CO.
TRADITONAL HOMEMADE GELATO
MsMsheireb Downtown Doha
Cloud&Co.

Cloud&Co.
Cloud&Co.
Cloud&Co.

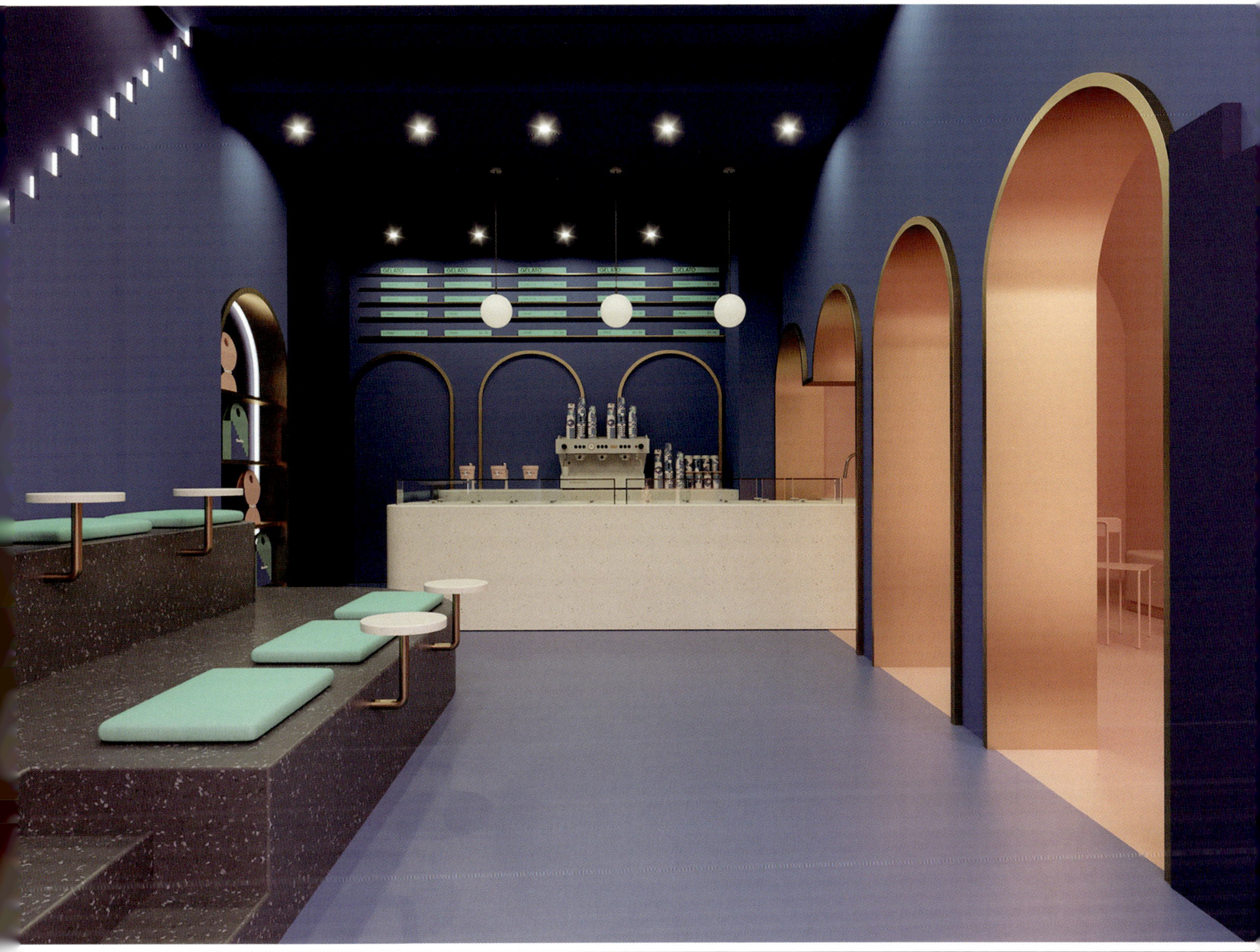

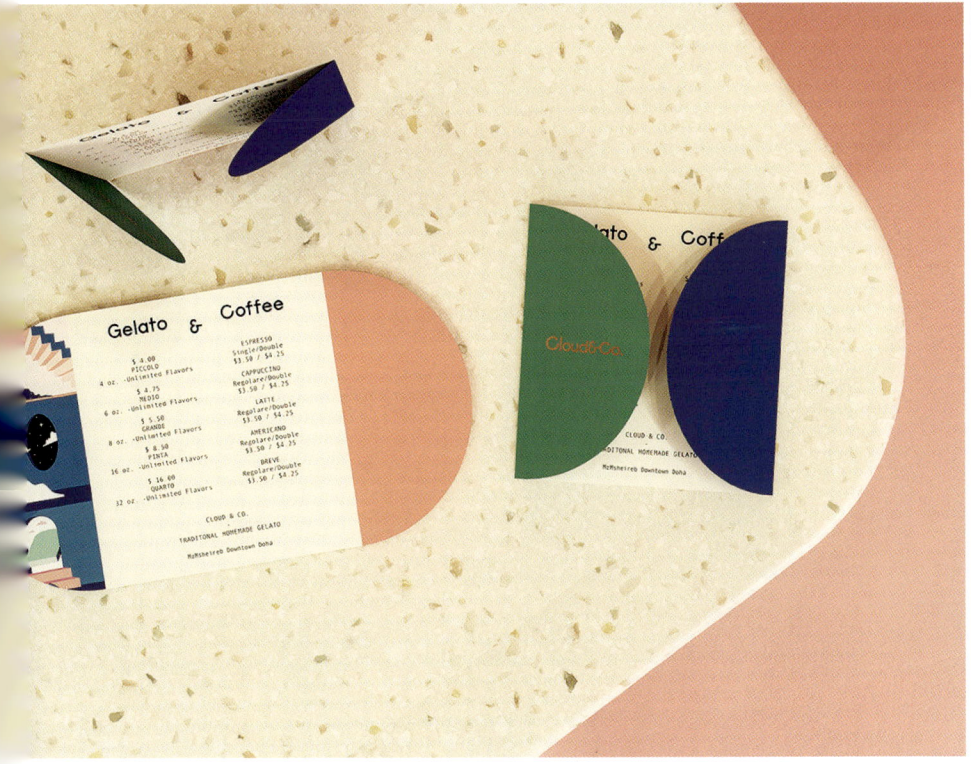

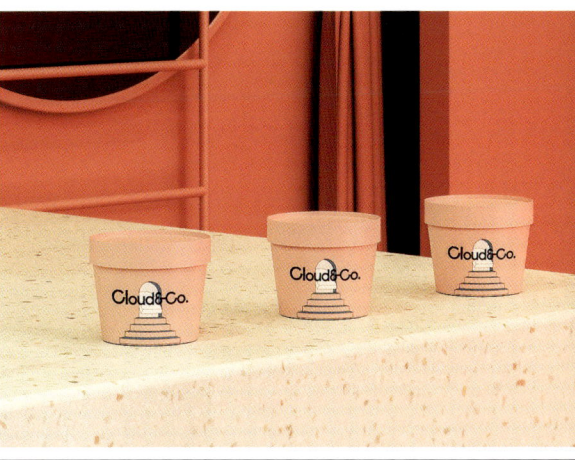

CIAM products perfectly fit into the architectural context of the venue. The central counter is the linking element between two distinct areas: the totally pink customer area, where an imaginative and playful world is created by incorporating elements that encourage interaction. On the other side, a totally blue environment creates a mysterious and dreamy scene. The supply consists of two pozzetti counters, placed in continuity and equipped with 16 carapines inserted flush with the worktop and provided with a handle in Glacer White Corian. The worktop, front paneling and counter sides are made of Corian Whitecap in a matt finish. The operator side was epoxy powder coated in traffic white RAL 9016 color in a matt finish.

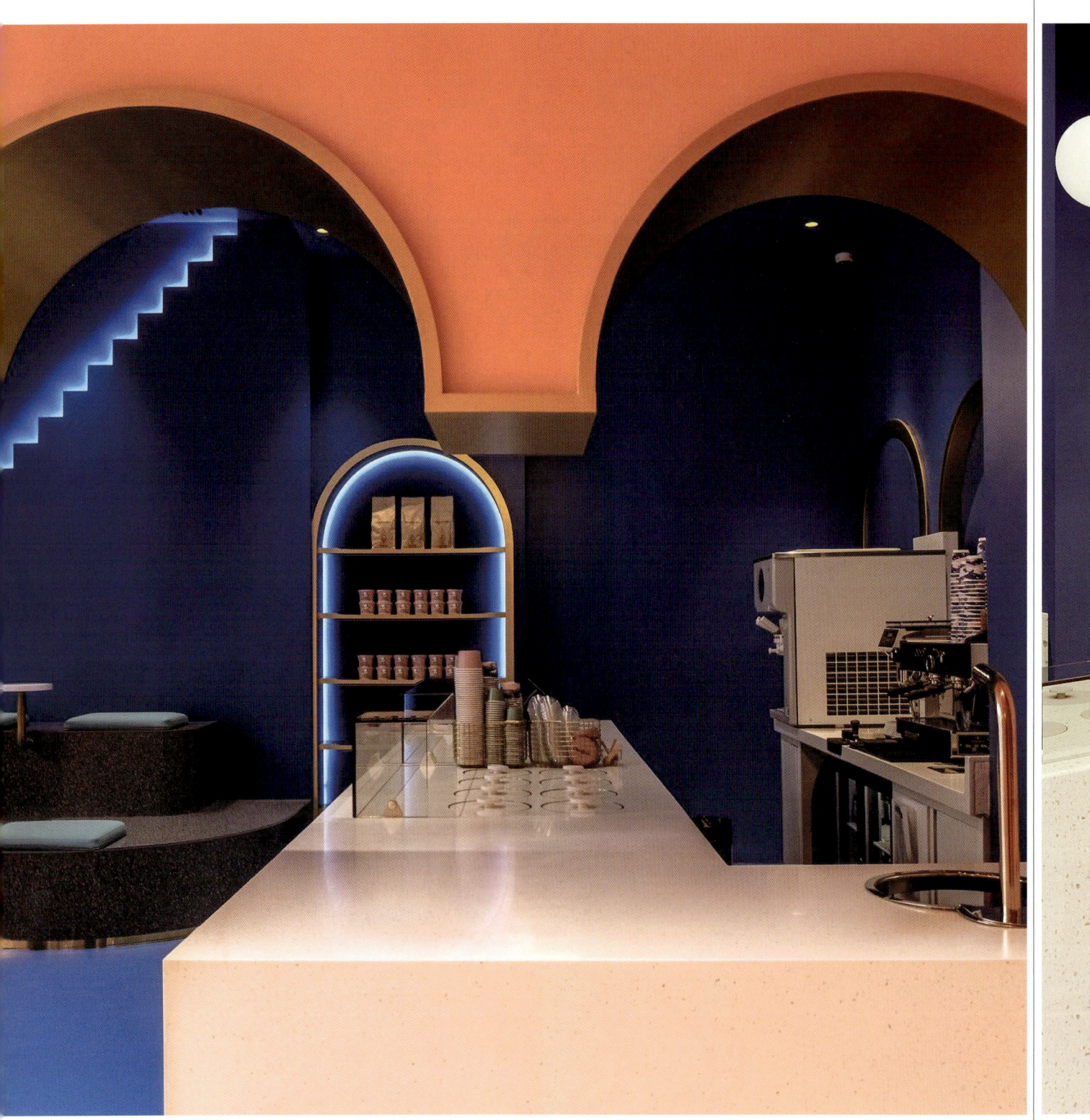

The central counter is the linking element between two distinct areas: the totally pink customer area, where an imaginative and playful world is created by incorporating elements that encourage interaction. Thier best selling dishes are Large Gelato, Macchiato, Gathering Box - 7 Medium Size and Brownie Sundae, although they have a variety of dishes and meals to choose from like Gelato, Cafe, Desserts.

▶Pasticceria Faiella – Salerno: Via Nizza, 103, 84124 Salerno SA, Italy / +39 089 285 7478

Pasticceria Faiella
Salerno SA, Italy

Historical pastry shop of Salerno~Pasticceria Faiella – Salerno

Opened in 1984, Pasticceria Faiella is a historic pastry shop in Salerno, run by master pastry chef Giuseppe Faiella, who took over an old Salerno pastry shop. The confectionery repertoire of Pasticceria Faiella embraces the entire culture of desserts from Salerno, Naples and, finally, Sicily. They range from fresh croissants to cakes, from shortcrust pastry to frozen desserts. The production is extremely varied and is characterized by the long experience of Giuseppe Faiella, whose name is particularly renowned in Salerno.

Traditional desserts, of course, as well as being inevitable are also highly appreciated. Worth trying are the babà (small or large), the puff pastry, the pastiera, the zeppole di San Giuseppe… and then millefeuille, cakes, pies. In short, Pasticceria Faiella is truly a pastry shop extremely rich in delicacies. Finally, the raw materials used for the desserts are top quality, always very fresh: this is one of the secrets of the pastry art of Giuseppe Faiella, who manages the place together with his son Alessio.

Pasticceria Faiella is highly appreciated for its traditional desserts. The babà, puff pastry and puff pastry, santarosa and pastiere are complemented by what is one of the hallmarks of the restaurant: puff pastry-babà with English soup. Pasticceria Faiella is not only traditional sweets from Campania: the cakes and semifreddos with exotic fruits (mango, papaya, coconut, etc.) are also worth trying, as well as cannoli, cassata and cassatine, typical Sicilian sweets.

Project details

▶ **Design:** Studio futura
▶ **Homepage:** http://byfutura.com
▶ **Area:** 160 m2
▶ **Location:** Via Nizza, 103, 84124 Salerno SA, Italy
▶ **Photography:** Bernhard Winkelmann

About us

In a world full of dull brands that look alike, the ones that stand out and create powerful relationships are the ones that succeed and last. Futura is a creative studio based in Mexico City, founded in 2008, with the goal of transforming traditional design forms and trying to change the way design is developed and consumed globally.

Our work is the result of constant experimentation, we believe in creativity above all else and how it can be translated not only in images, but also in objects and spaces. The essence of Futura is in making perfection meet chaos. We create and revamp brands that stand out to create meaningful relationships with people, empower businesses and resonate around the world. We believe that the only way of being relevant nowadays is to create authentic brands that stand out locally and globally connecting and developing lasting impact with people.

Ritz PARIS
LE COMPTOIR

Ritz
LE COMPTOIR

Ritz
PARIS
PÂTE À TARTINER
220 g
7.76 oz
HAZELNUT & COCOA SPREAD
FABRIQUÉ EN FRANCE

Ritz
PARIS
210 g
FRAMBOISE
7.4 oz
RASPBERRY
Préparation à base de fruits
FABRIQUÉ EN FRANCE

Ritz
PARIS

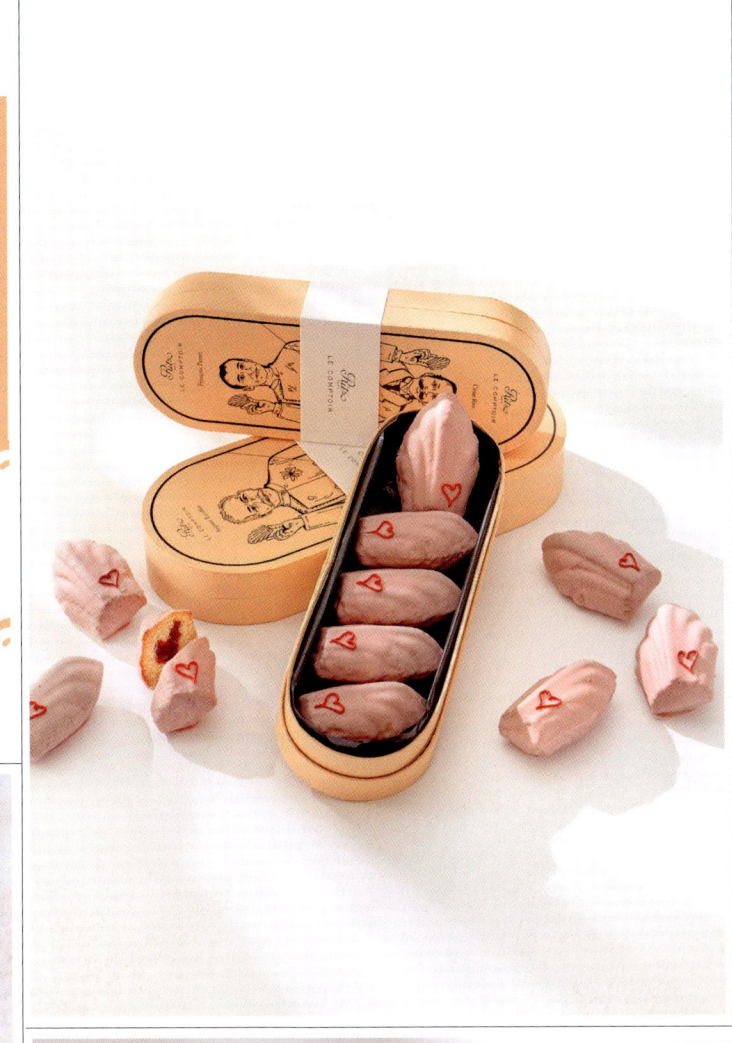

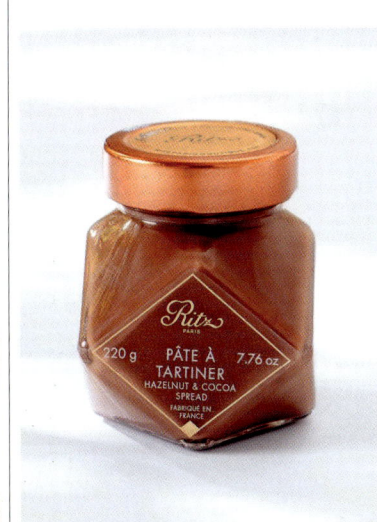

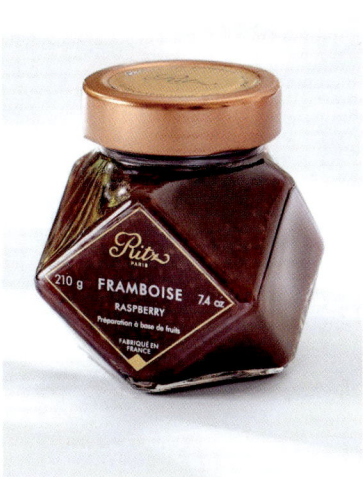

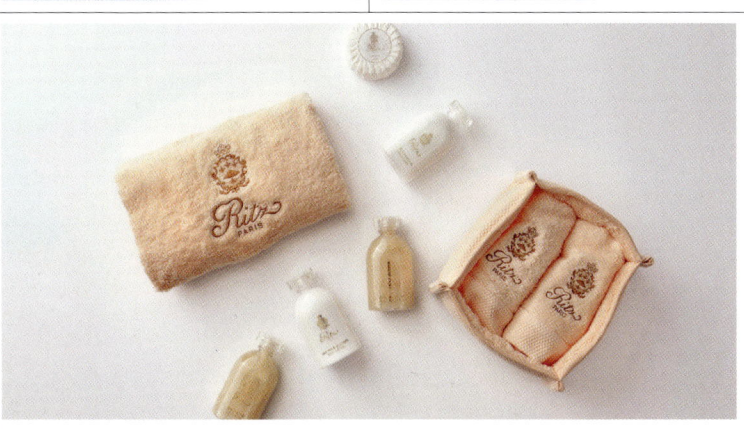

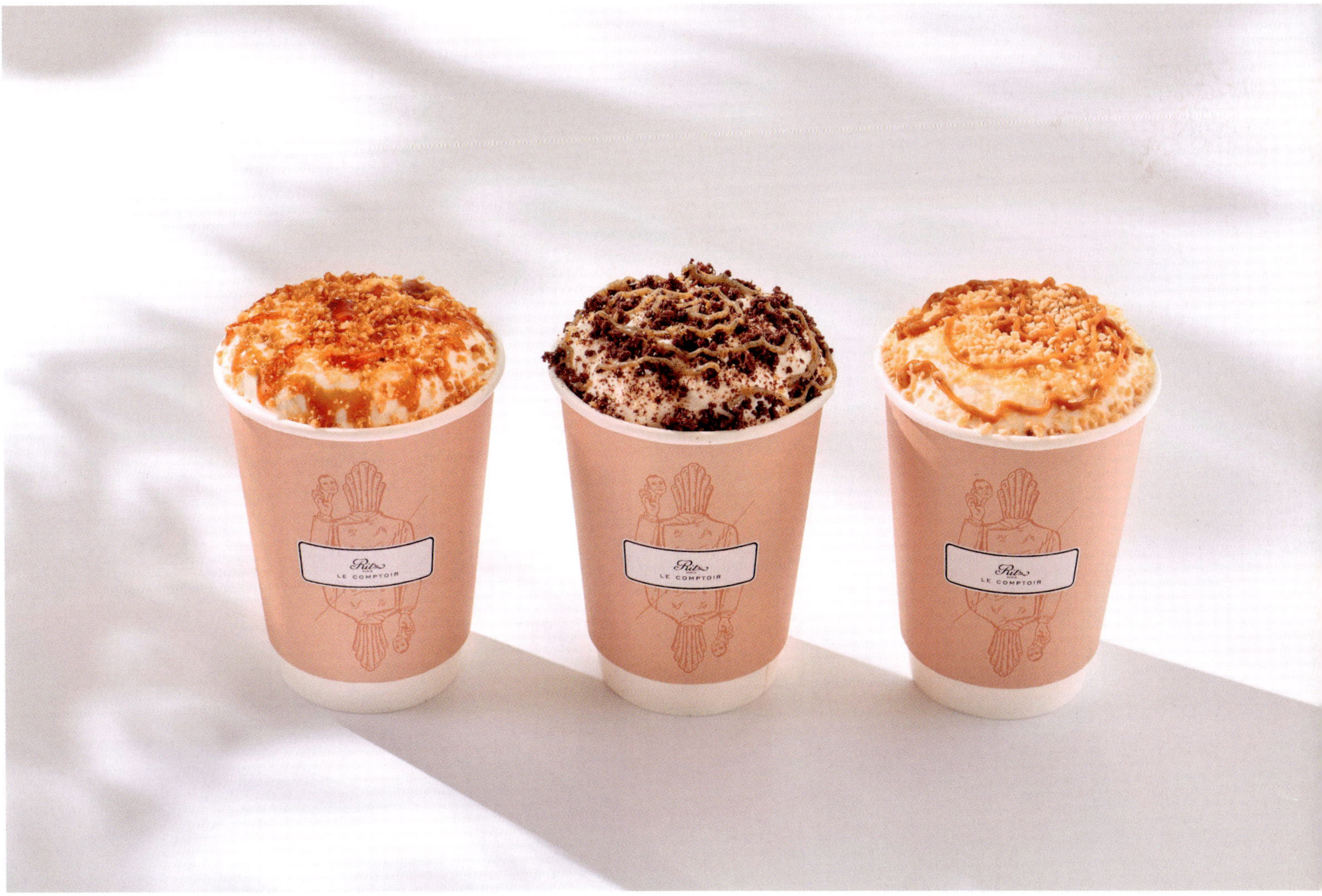
Ritz
LE COMPTOIR
Ritz
LE COMPTOIR
Ritz
LE COMPTOIR

A contemporary and elegant lifestyle space.
The Ritz Paris has welcomed the world's most illustrious guests and, more than a century after it first opened, its status as an icon of gracious living remains intact. The hotel is the eternal symbol of glamour, the avant-garde, and Paris as a literary capital. True to this DNA, the Ritz Paris today wishes to share with Parisians part of what makes it so marvelous. A veritable destination with a separate entrance at 38 Rue Cambon, Ritz Paris Le Comptoir fits seamlessly into the neighborhood and into everyday life for gourmets of every stripe. From morning to night, Ritz Paris Le Comptoir serves as a destination for neighborhood residents — professional and otherwise — to drop by for a delicate offering from an exceptional House. Available by click-and-collect, a concentration of François Perret's best pastries will delight gastronomes from all over. Clients may savor these sweet indulgences right at the counter, or pick them up before strolling through the city or attending a dinner — it's also possible to find a perfect dessert with a wine or champagne pairing.

A Ritzy urban annex.

Conceived as a welcoming, lively place that's rooted in everyday Parisian life, Ritz Paris Le Comptoir is nonetheless refined and true to the chic style of the hotel. At once functional and elegant, this showcase brims with noble materials, made-to-measure details, and plays on texture that echo its refined pastries and create a fresh sense of conviviality. China cabinets, Art Deco dessert trolleys, and pastry counters recall the pedestal tables of yesteryear while, behind a window, one spies the bustling ambiance of a "pastry lab." Subtle lighting from a spectacular chandelier recalls the generous curves of a madeleine, while sinuous alcoves are clad in gold, and a display has been hewn from a single block. There are naïve, almost childlike, illustrations and a large, gracious portrait of François Perret. For those wishing to sit down, an enveloping couch awaits, with the added comfort of being tucked behind beautifully worked screens, out of sight of onlookers in the street.

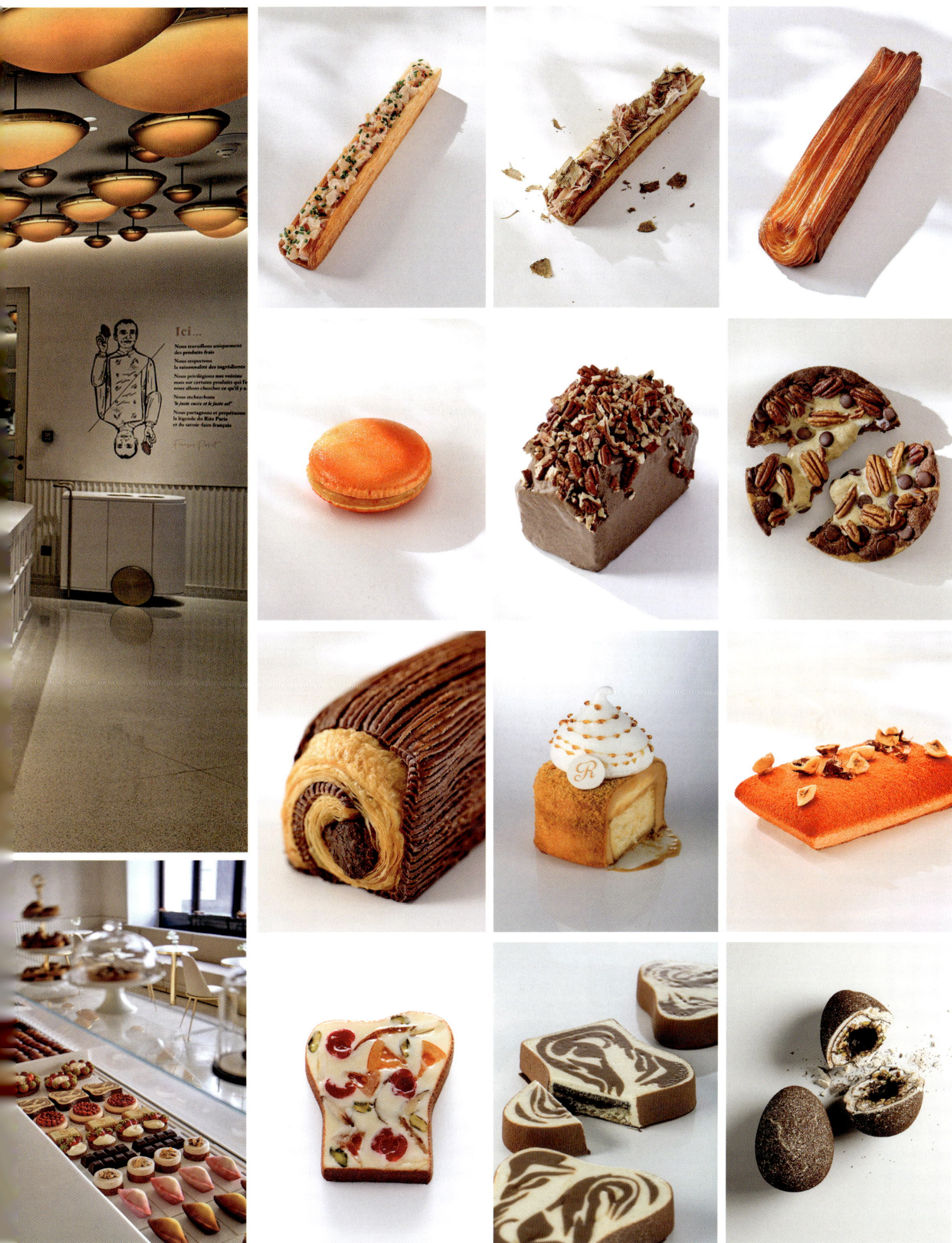

The quintessence of French excellence, the Ritz Paris inaugurates a new chapter in its history with the opening of the Ritz Paris Le Comptoir, the gourmet boutique by François Perret. A new destination that's open to all, this Ritzy urban annex is distinctive for its ambiance of conviviality, sharing and gourmandise.

About François Perret.
François Perret was named the World's Best Restaurant Pastry Chef in 2019, the same year his book "French Pastry at the Ritz Paris" was published by La Martinière, and with the success of the first season of "The Chef in a Truck" series on Netflix, his audience keeps growing daily. An abundance of recognition for the Chef, a native of Bourg-en-Bresse, who still vividly recalls the epiphany of his father's crème anglaise. Having worked in some of the best kitchen brigades in Paris - at the Lancaster, Le Meurice, the Four Seasons George V and the Shangri-La Hotel Paris, François Perret was granted carte blanche at the Ritz Paris to create the House's new pastry identity when it reopened in 2016.

With its bright, contemporary décor, Parisians as well as visiting gourmets will find a welcoming space in which to sample the crème de la crème of baked goodies. Here, foodies can experience first-hand the mastery of Perret's marvellous creations, which include a complete redressing of the iconic pain au chocolat, fluffy breads, small sandwiches and, of course.

Traditional desserts, of course, as well as being inevitable are also highly appreciated. Worth trying are the babà (small or large), the puff pastry, the pastiera, the zeppole di San Giuseppe... and then millefeuille, cakes, pies. In short, Pasticceria Faiella is truly a pastry shop extremely rich in delicacies.

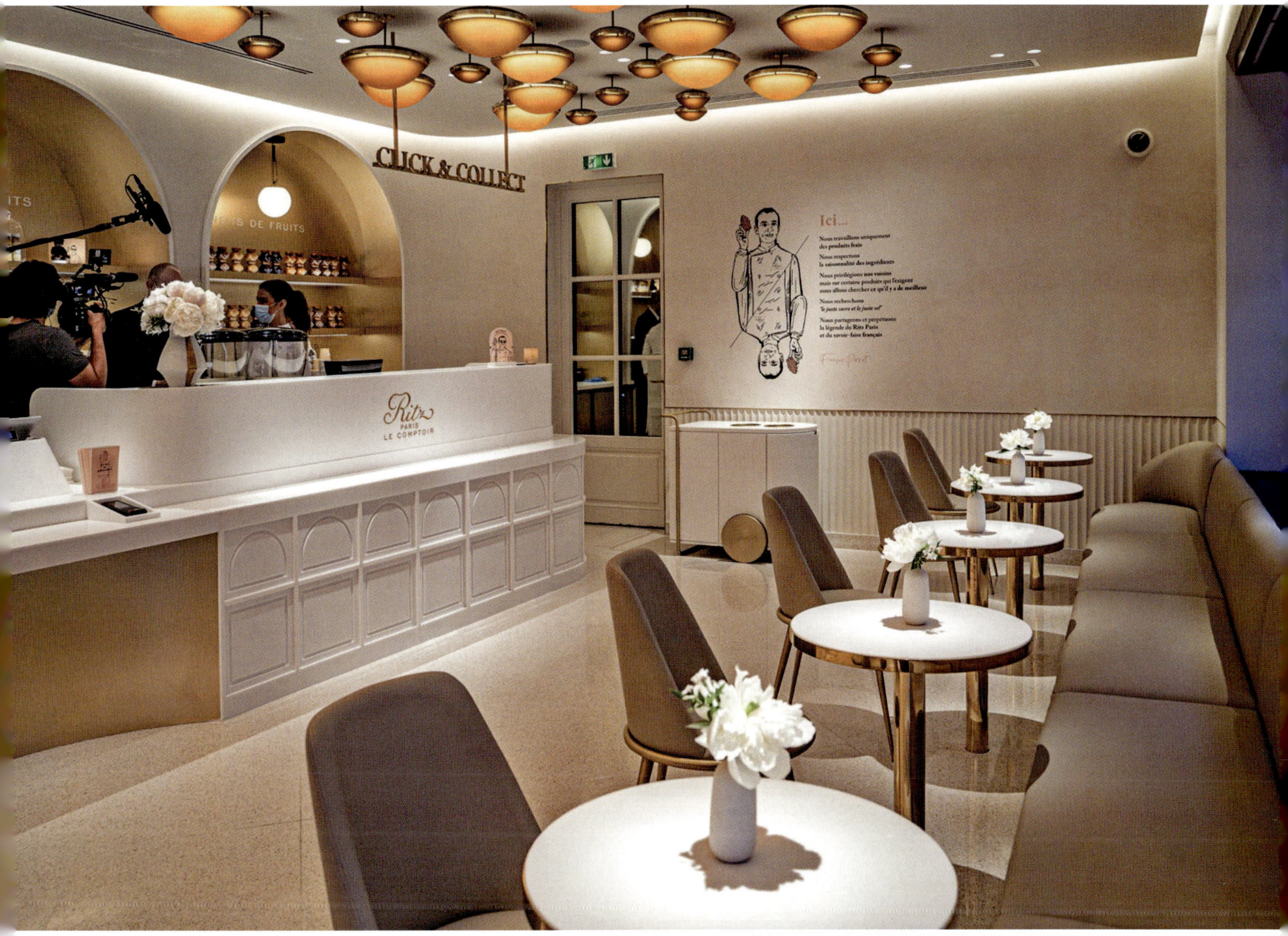
CLICK & COLLECT
Ritz
PARIS
LE COMPTOIR
Ici...
Nous travaillons uniquement
des produits frais
Nous respectons
la saisonnalité des ingrédients
Nous privilégions nos voisins
mais sur certains produits qui l'exigent
nous allons chercher ce qu'il y a de meilleur
Nous recherchons
"le juste sucre et le juste sel"
Nous partageons et perpétuons
la légende du Ritz Paris
et du savoir-faire français

Ici...

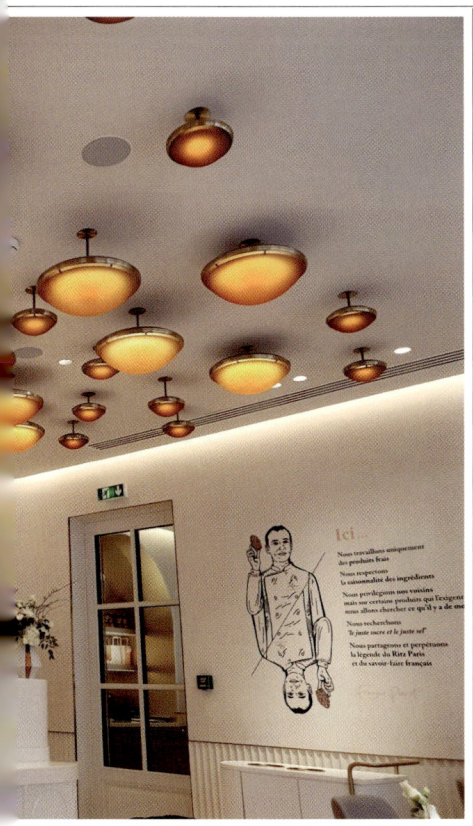

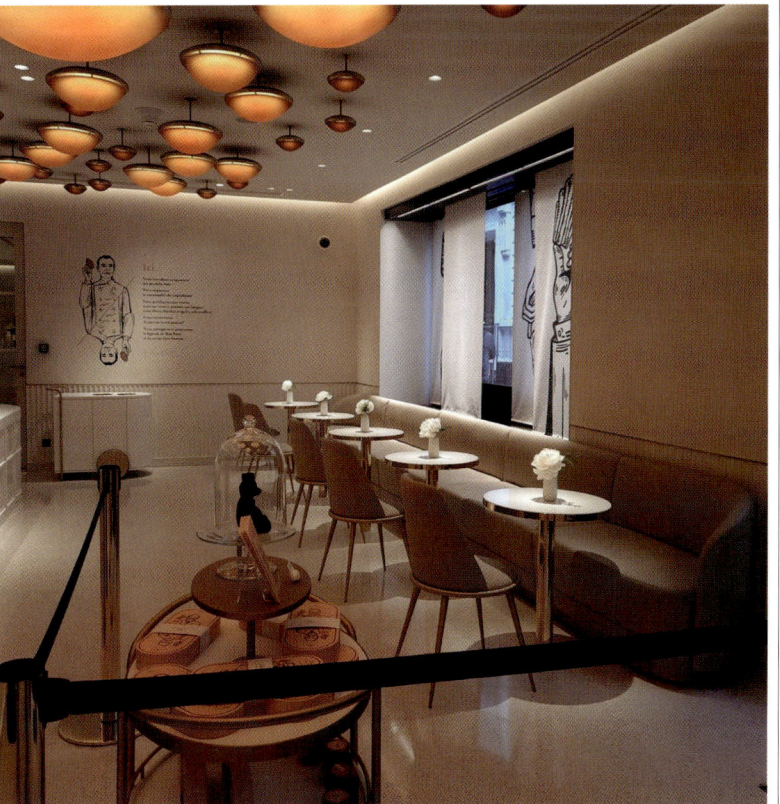

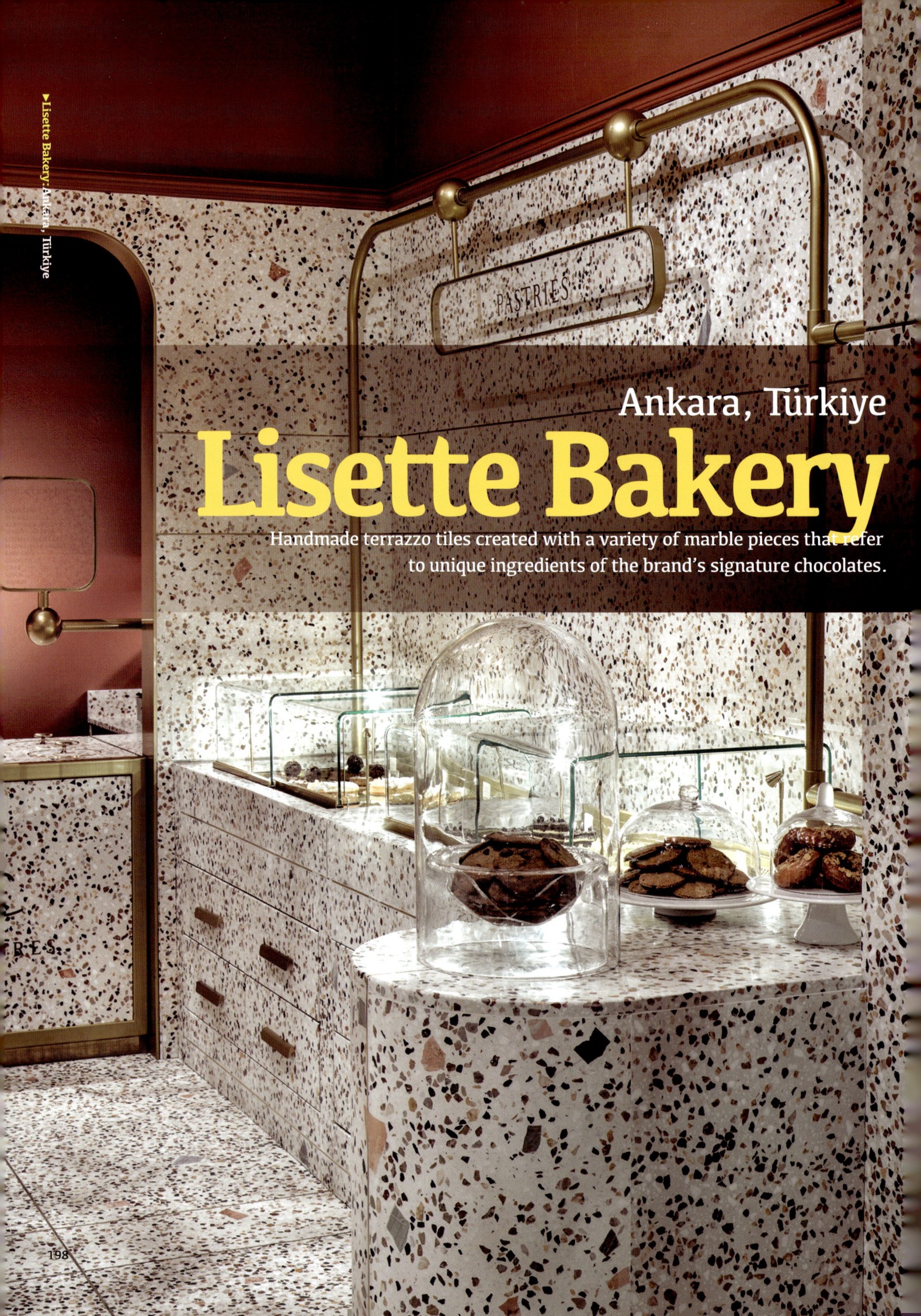

Lisette Bakery

Ankara, Türkiye

Handmade terrazzo tiles created with a variety of marble pieces that refer to unique ingredients of the brand's signature chocolates.

Nēowe delivered a luxurious and bespoke retail space showcasing Lisette's finely decorated chocolates, cakes, and icecreams like the treasures found in a jeweler's shop. The color palette inspired by Lisette's brand identity can be seen reflected in the materials and details inside the space. Alongside the custom terrazzo surfaces, reflective installations covering the ceiling, brass, and stained glass used on the front façade emphasize the sophisticated, elegant, and characteristic feeling of the brand that reflects into space. Inspired by the signature half dome chocolates of the brand, Lisette's 44 sqm interior space, with its highly detailed manufacturing and material mixture, offers a memorable experience for the visitors.

The handmade terrazzo tiles were created with a variety of marble pieces that reference the unique ingredients of the brand's signature chocolates. The display units, floors, and walls are offering a seamless transition between the surfaces, and continuity in the space is provided with one-piece brass strips used at the joints of the blocks. Specially designed circular units located in the center of the glass lanterns force the boundaries of the construction without breaking the rules.

Brass frames that emphasize architectural character and the front facade, weaved from stained glass, create a different atmosphere that breaks the intense flow of the public area, while the guests are welcomed together with the brand's "Guilty Pleasures" label created from illuminated onyx marble letters hidden on the floor. In the interior space, where Art Deco elements and design approaches are dominant, brass and terrazzo materials provide integrity by intertwining the whole volume and creating a warm, detailed, multi-layered, and value-added space design.

Project details

- **Design:** Neowe design studio
- **Homepage:** www.neowe.com.tr
- **Area:** 44 m2
- **Location:** Ankara, Türkiye
- **Photography:** Ibrahim Ozbunar

About us

Neowe was established in 2017 by co-founders Nil Emiroğlu, F. Sertaç Kılıç and Ferhat Özkan. Neowe is dedicated to the highest levels of details and creative expressions by focusing on unique and highly customized interior spaces. Team exploration of fashion forward and timeless designs creates a seamless juxtaposition between contemporary and classic, graphic and instinctual, structural and organic disciplines.

Neowe's design is driven by a deep passion of design and realization of projects as coherent as possible. Team's commitment to relationships with their clients, fabricators and fellow designers are keys to successful projects. Neowe's main goal is to embody integrity, originality and love of desing in each of its projects togetger with a shared sense of values and vision among all stakeholders.

facade

The geometries, materials, and pastel colors that have become a part of the design, emphasize the brand identity of Lisette while providing an unforgettable experience for the customers, the texture surrounding the interior offers visitors a chocolate box illusion that is carved out of a single piece of terrazzo cube.

The terracotta-colored walls create an impressive background for the installation consisting of brass flakes illuminated by the spotlights, while the ice cream unit and brass jambs covered with terrazzo plates reflect the designers' essential attention to space. Custom-made tables, specially designed by Neowe, mediate the transportation of materials and geometries used in the interior with the combination of brass and terrazzo while the seating that is signed "Lisette" emphasizes the brand's aesthetic and timeless design approach. The brief was to showcase finely decorated sophisticated chocolates, cakes, and ice cream with a display almost like a glass cabinet in a jeweler's shop.

longitudinal section

cross section

205

The main challenge was the space limit in this project. A wide variety of product showcase units, a background preparation area, and an open display window facing the facade were all arranged in a 44 sqm space. Every material and equipment was produced custom for this project in order to solve area limitations. Even manufacturing drawings of industrial cooling equipment were controlled by the design team in every phase. The main design solution was using one main material in-ground, wall, and all surface coverings, which was handmade terrazzo.

plan

Famiglia Ice Cream
Córdoba Province, Argentina

The ceiling and the incorporation of vegetation to contribute to a harmonious environment.

Famiglia is much more than an ice cream store, it is a meeting point that makes people talk, firstly because of the richness of the ice cream and secondly because of everything that the brand generates. Based on these premises, we sought to innovate the way in which the space was conceived, inspired firstly by the creaminess of the ice cream and secondly by the corporate elements that the brand as such had. In the branches, the item and the tools were the same, but each design was adapted to the particularity of each location. The study designed, built and managed the work in its entirety, including the tests and experimentations that the ceiling carried. The material exploration was the main challenge, to achieve from the material a unique, memorable sensation, like eating an ice cream here. To achieve this result, we work with various materials until we find a mesh fabric supported on a net and translucent, with a very cautious art direction, we seek to form curves, waves, spirals and organic movements.

The distribution in the premises seeks to make the most of its uses in a clear and transparent manner. For this, the study takes advantage of the existing conditions of the premises, achieving a concise communication between the interior and the exterior, accompanying the gesture with continuous furniture, the incorporation of the interior landscape and the present colorimetry. The equipment of both ice cream parlors was specially designed according to the needs of each one. For example, the tables proposed for the use of the public of ice cream parlors seek to adapt to the distribution of each branch and the morphology of each local. As well as the design of the curved lines that complement and link between the artistic finish on the wall, the mirrors, the ceiling and the incorporation of vegetation to contribute to a harmonious environment.

Project details

- ▶**Design:** SET Ideas
- ▶**Homepage:** www.neowe.com.tr
- ▶**Area:** 50 m²
- ▶**Location:** Córdoba Province, Argentina
- ▶**Photography:** Gonzalo Viramonte

About us

Optimizing architecture may seem like an overly ambitious dream, elusive. But what happens when a team pushes him towards that dream? Three years ago we began to bring SET to life. Three years ago we began to bring to life the idea of small units optimized to the maximum to take advantage of every square inch. For three years we have been pursuing the dream of a much simpler way of inhabiting and enjoying spaces. And today, three years later, we are proud to contribute every day so that this dream of an architecture that surpasses us and our environment is closer and closer.

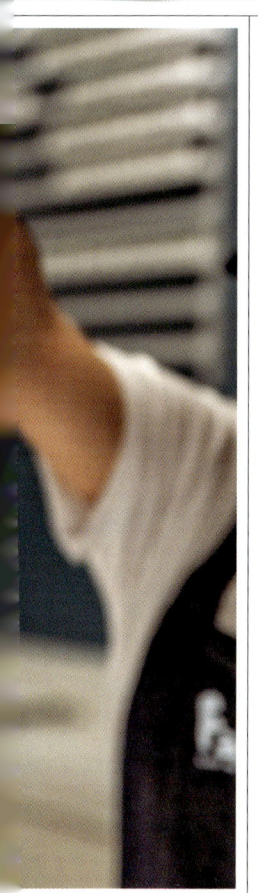

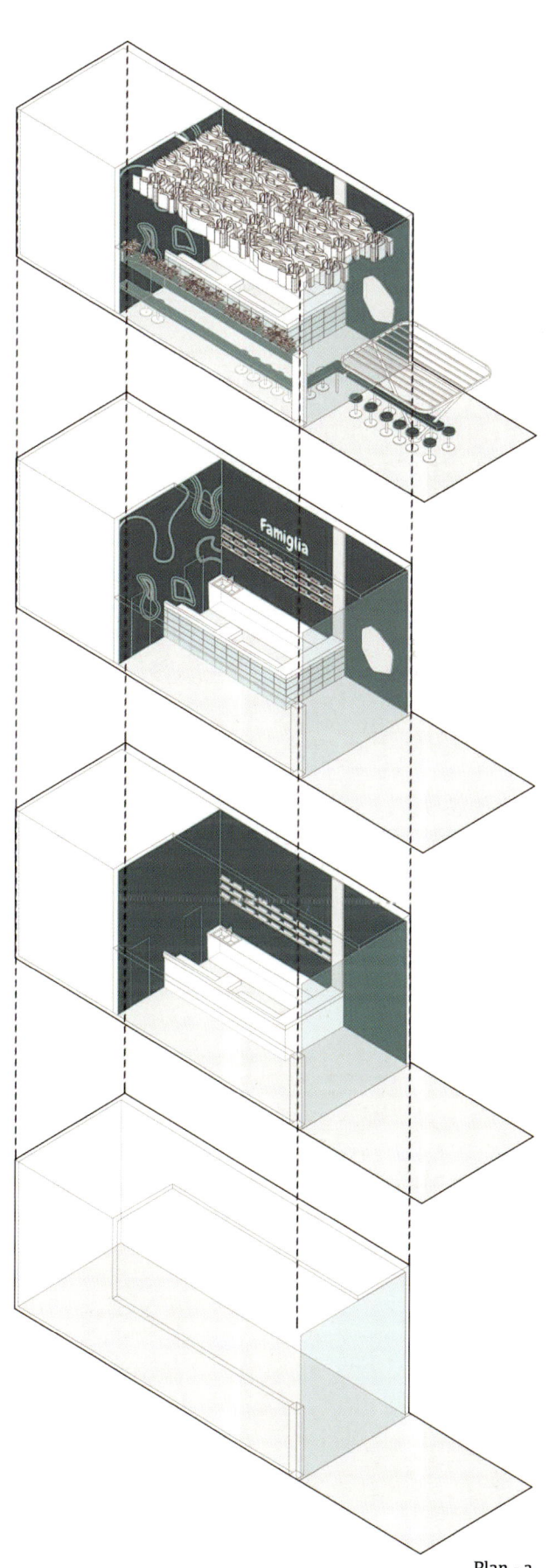

Plan - a

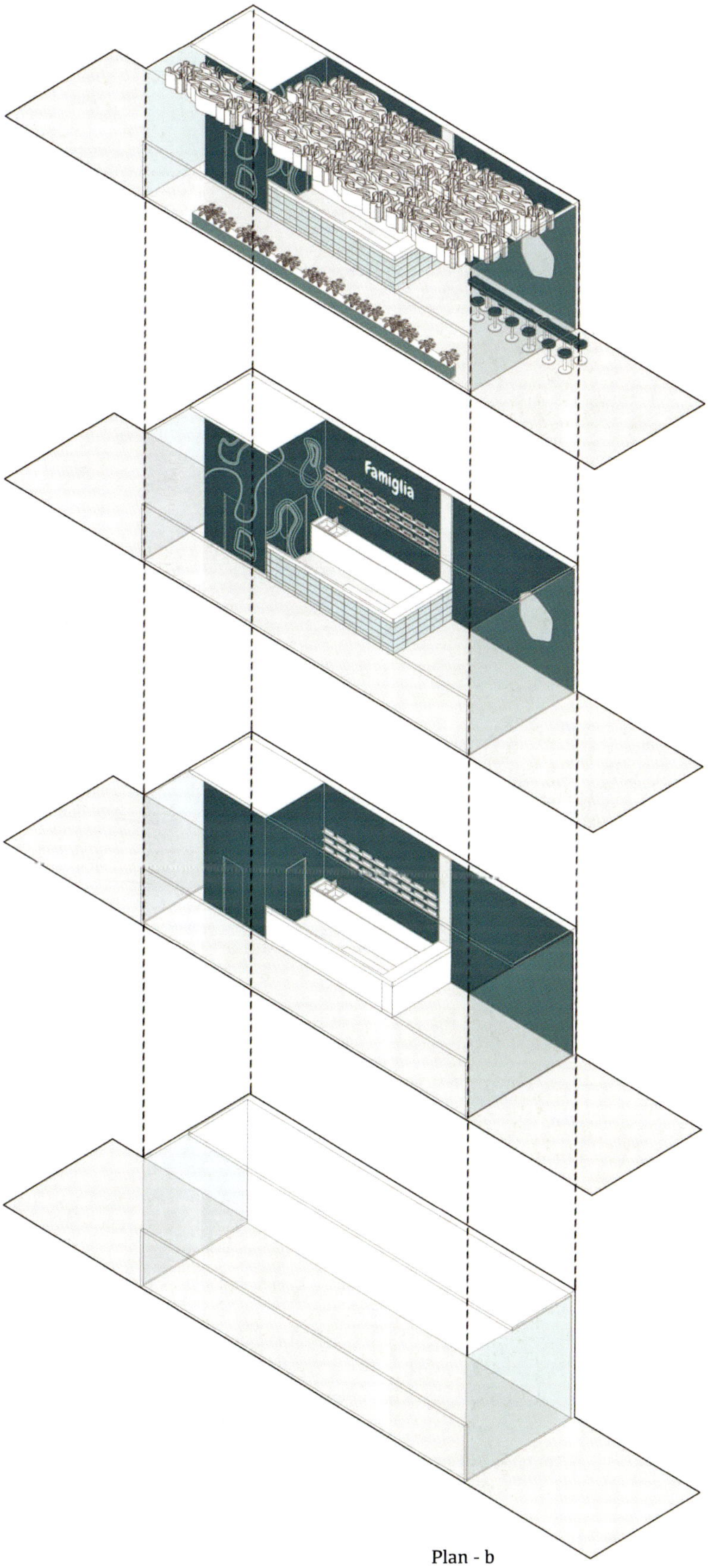

Plan - b

217

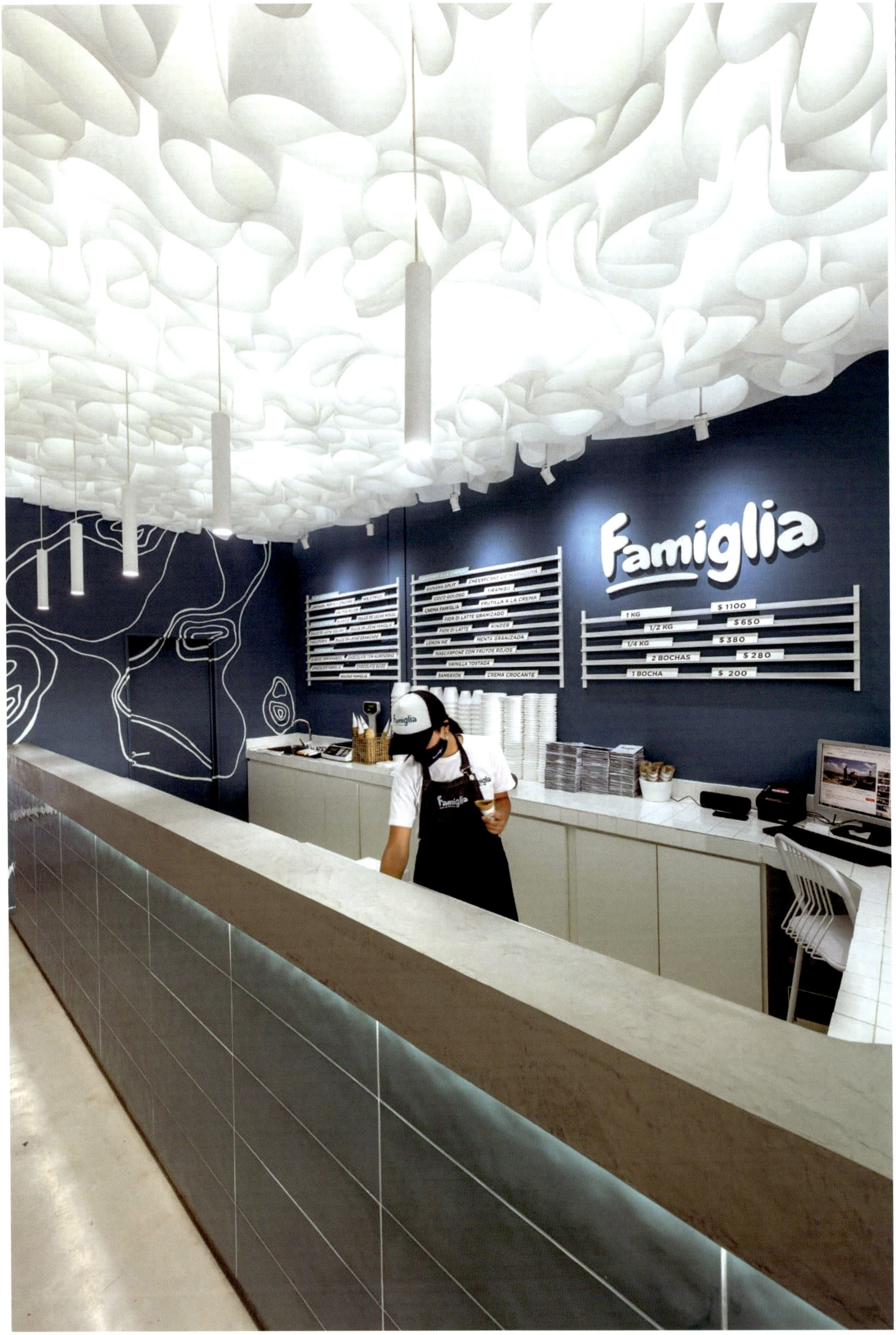
Famiglia
1 KG
$ 1100
1/2 KG
$650
1/4 KG
$380
2 BOCHAS
$ 280
1 BOCHA
$ 200

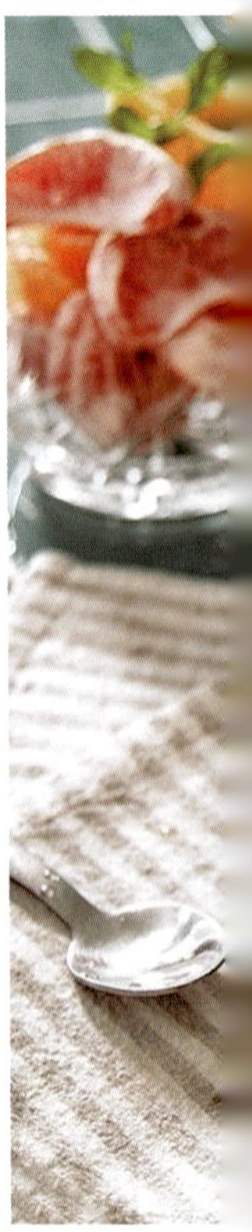

Ofelè Pasticceria
Lavena ponte Tresa VA, Italy
The neoclassical style reinterpreted in a contemporary key-Ofelè Pasticceria – Lavena Ponte Tresa (VA)

The Ofelè project was born in 2018 from the meeting, at the international school of Italian cuisine ALMA in Colorno, of the Pastry Chef Linda Chirico and the Architect Simone Colombo. From that meeting comes a creative synergy and a common vision that leads, in December 2019, to the opening of the first Ofelè store in Porto Ceresio (VA) and today, July 2022, of the second store in Lavena Ponte Tresa.

The architectural project features a neoclassical style reinterpreted in a contemporary key, where the Colors and the Materials Palette tell of a journey into the wonderful world of Ofelè. A world made of design, detail, care, patience and passion fueled only by the positive energy of a smile. The furnishings, the materials, the lights, the music are designed to be in harmony with the shapes and colors of the creations of Pastry Chef Linda Chirico. Antique pink, oxidized brass, ribbed glass, walnut wood, calacatta marble, grit, velvet and tulips, many tulips.

Project details

- **Design:** AFA Arredamenti, Arch. Simone Colombo
- **Homepage:** https://www.afa-arredamenti.com
- **Area:** 160 m²
- **Location:** Lavena ponte Tresa VA, Italy
- **Photo Image:** AFA Arredamenti

About us

Afa was founded in 1968 by the Francolini brothers. We are very careful in maintaining the artisanship of our origin, but today the precision and mastery of our techniques are even more important, thanks to the use of modern and sophisticated machines.

AFA designs, produces, and installs: bespoke furniture, boiserie, moquette and parquetry, doors and frames, fabric accessories, upholstery, curtains, lightings, and accessories. We follow the entire process that allow the realization of your locations: from material research to design and realization of furniture. Our production cycle covers all areas: wood, metals, upholstery, fabric and lighting. In this way nothing is left to chance and we can control everything: design, materials, and execution quality.

OFELÈ
CAFFÈ E PASTICCERIA

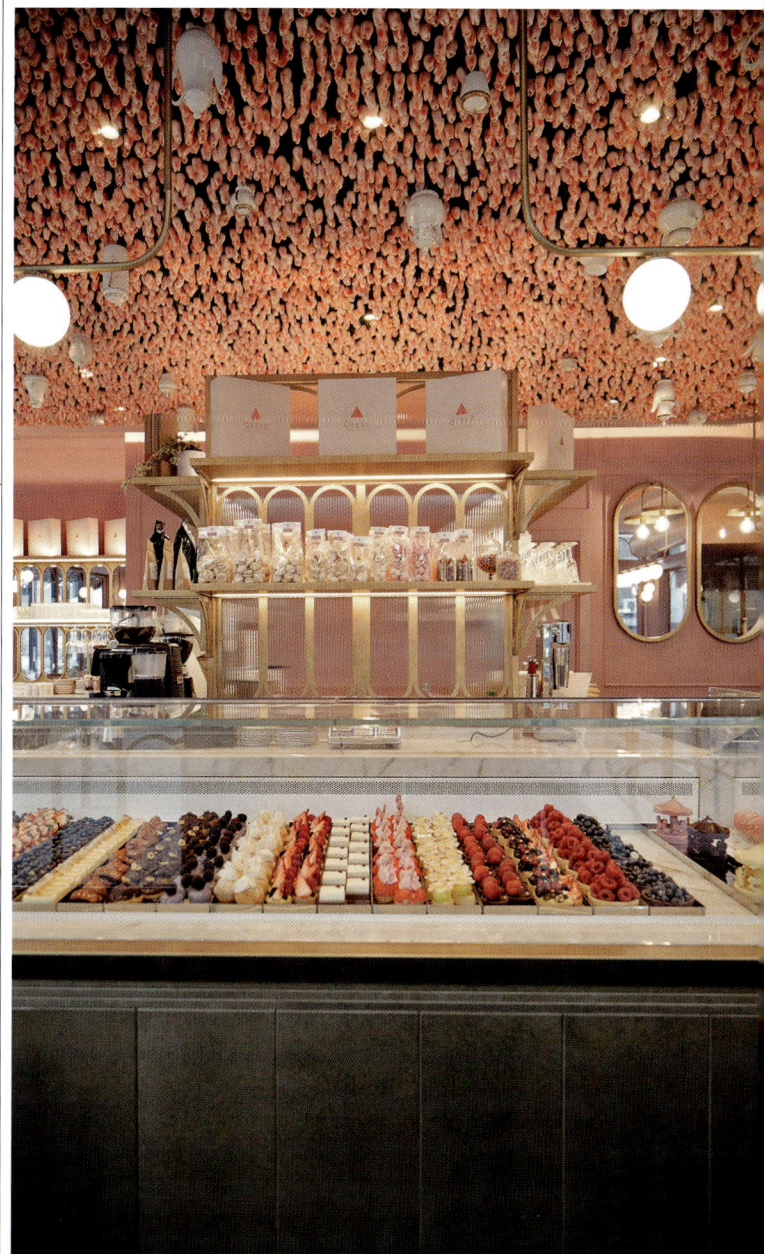

225

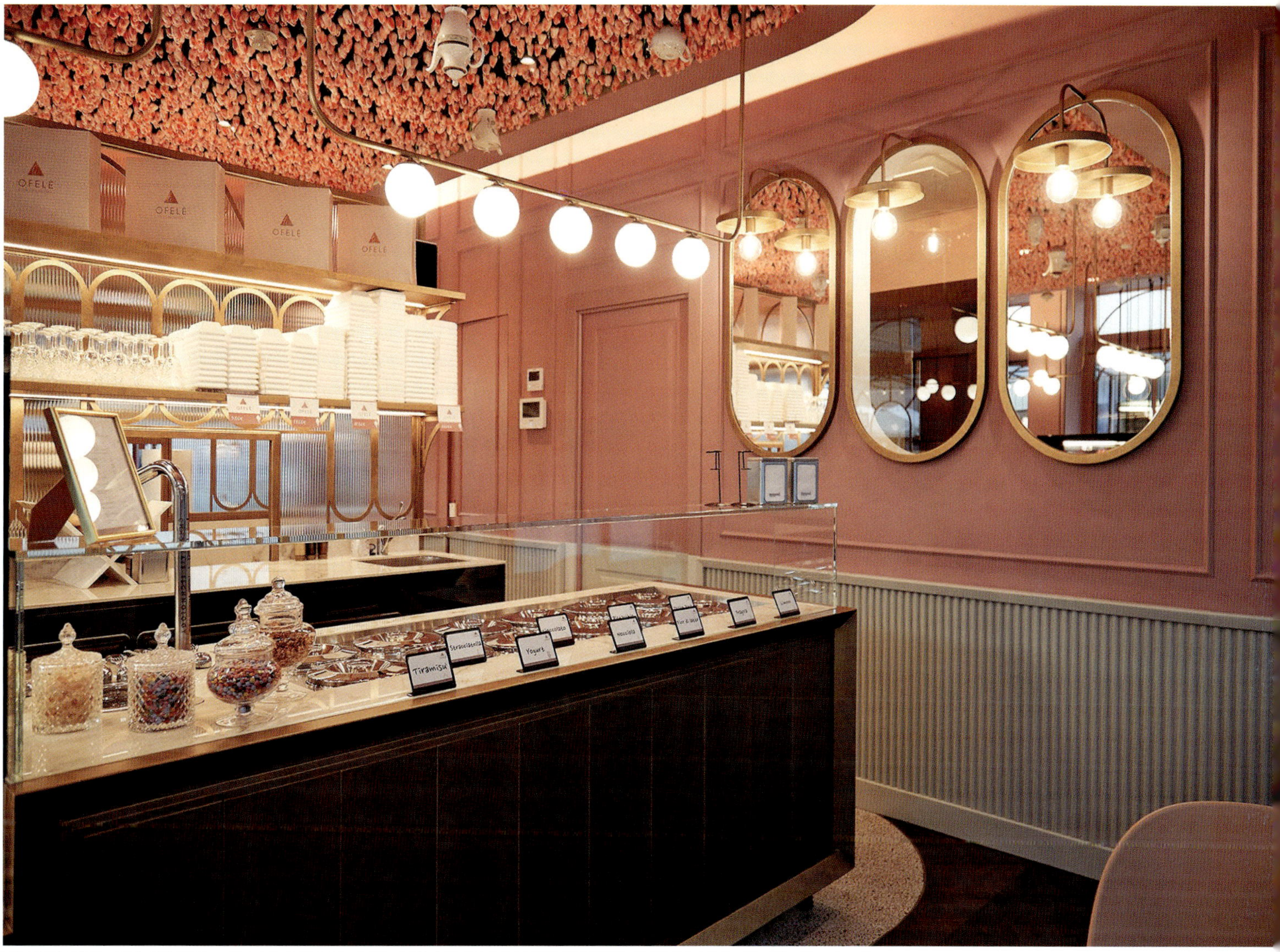

▶Picnic Salad & Bakery: Perämiehenkatu 6, 00150 Helsinki, Finland / +358 40 7309612

Helsinki, Finland
Picnic Salad & Bakery
Fresh look for a new brand position - Picnic Salad & Bakery

The business target: With around 40 locations across Finland, Picnic is the go-to restaurant chain for quick meals and favourite cafeteria products. They were embarking on a fundamental strategic repositioning with the focus on good-conscience fast food. However, Picnic's old brand identity and interior concept told a different story. They invited Kuudes to perform a major overhaul of the Picnic look in order to freshen the image customers have about the brand.

Our solution(Fresh ingredients first): To communicate the freshness of Picnic products, we let their ingredients take center stage. Boxes of potatoes and bowls of lettuce are not hidden behind the counter – instead, they are proudly presented, visible to the customer. We also created a pattern with organic shapes, inspired by Picnic products and ingredients like baguettes, apples and tomatoes. The illustrations are used everywhere from restaurant walls to take away packaging to create a consistent look.

Eat your colours: Colours play a crucial role in the new customer experience. The natural and organic palette embodies the energy and freshness of Picnic ingredients. But it has a highly functional purpose too. The bright brick red guides the customer to the fast lane with take-away options, self-service, and the 'PicUp' station for pre-orders. The refreshed look is especially important when innovating new products suitable for the rise in takeaway dining. With clearly marked taste options and upgraded usability, the packaging is now not only inviting to the eye and taste buds but also convenient in its grab-and-go nature. The proudly biodegradable packaging range makes sure you can enjoy your meal with a clear conscience, for yourself and the planet.

Project details

▶ **Design:** Kuudes
▶ **Homepage:** https://kuudes.com
▶ **Area:** 40 m²
▶ **Location:** Helsinki, Finland
▶ **Interior Image:** Henri Vogt
▶ **Packaging images:** Tomas Olsen of Studio Fotonokka

About us

Kuudes is your partner in identifying growth opportunities. We create new commercial concepts and validate their market demand with our insight, foresight and strategy capabilities. We then bring the concept to life with world-class digital and physical design.

We believe foresight is crucial in a constantly changing world. Together with prominent innovators, we developed the Kuudes Next foresight model to enrich your strategy and innovation work over a timespan of 3-5 years. With previews of disruptive consumer phenomena, you are able to evaluate vital strategic decisions and the timing of innovations. We partner in designing strategies and commercial concepts, as well as in creating new product and service innovations. We crystallise customer-driven brand strategies and translate them into visual identities and engaging design. We design digital and physical services as well as retail environments to create consistent brand experiences.

Discover opportunities for growth by bringing in-depth customer insight and creative thinking into business development.

picnic

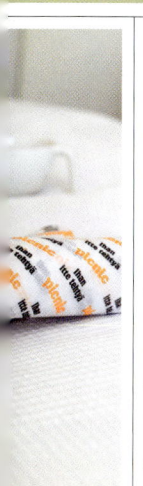

Results: "Working with Kuudes is always a pleasure. Their approach is systemic and pragmatic, always keeping both consumer insight and business objectives in mind. As the scope of transformation is so wide in this case, Kuudes was able to utilize their capabilities to the fullest. This gave a special flavor to the collaboration as we had the opportunity to cover all consumer touchpoints. The end result is on strategy, distinctive and interesting. We continue being very satisfied with collaborating with Kuudes and the talents working there."

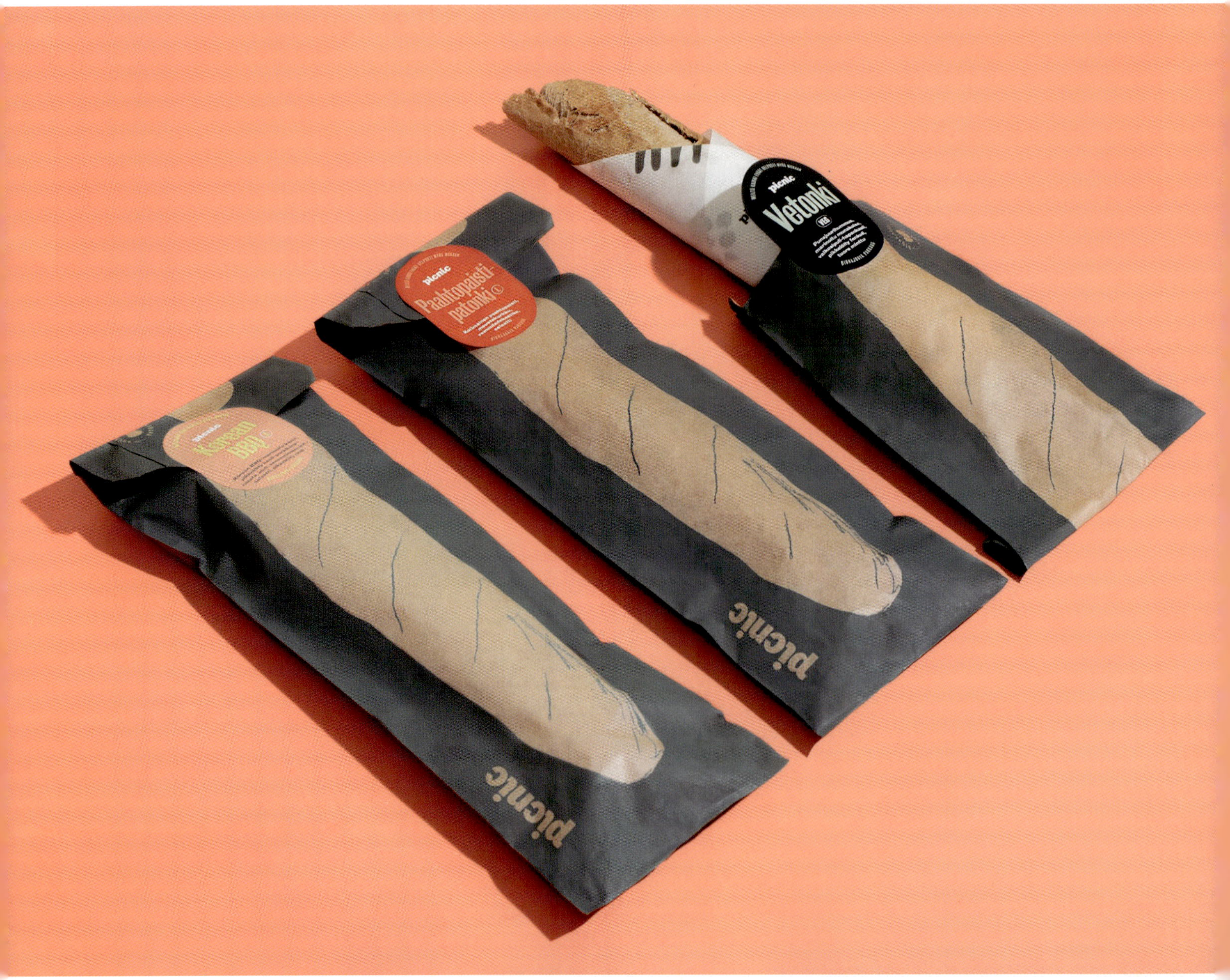
picnic
Korean BBQ
picnic
Paahtopaisti-patonki
picnic
Vetonki
picnic
picnic

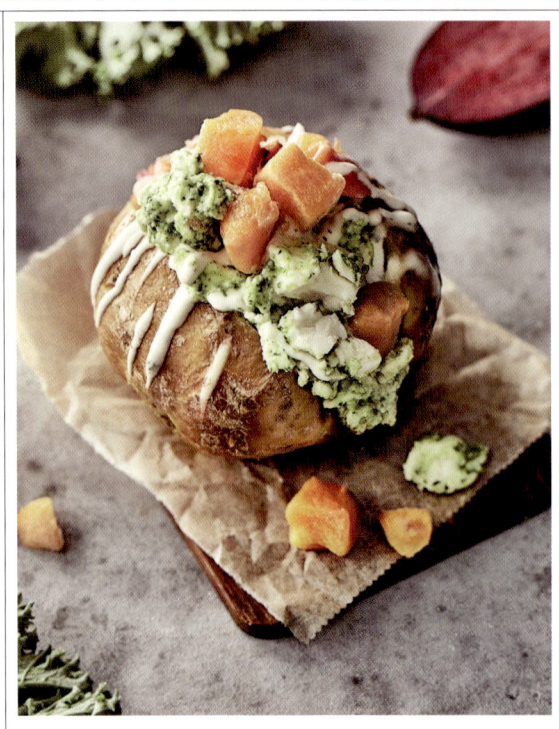

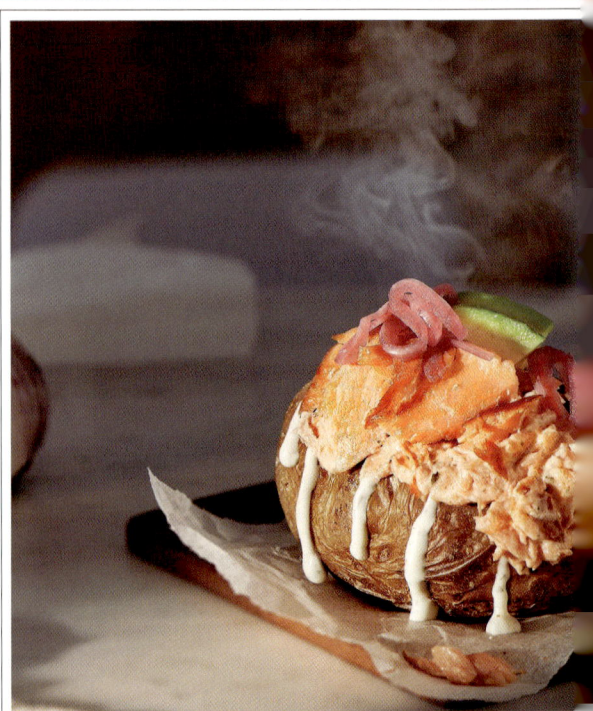

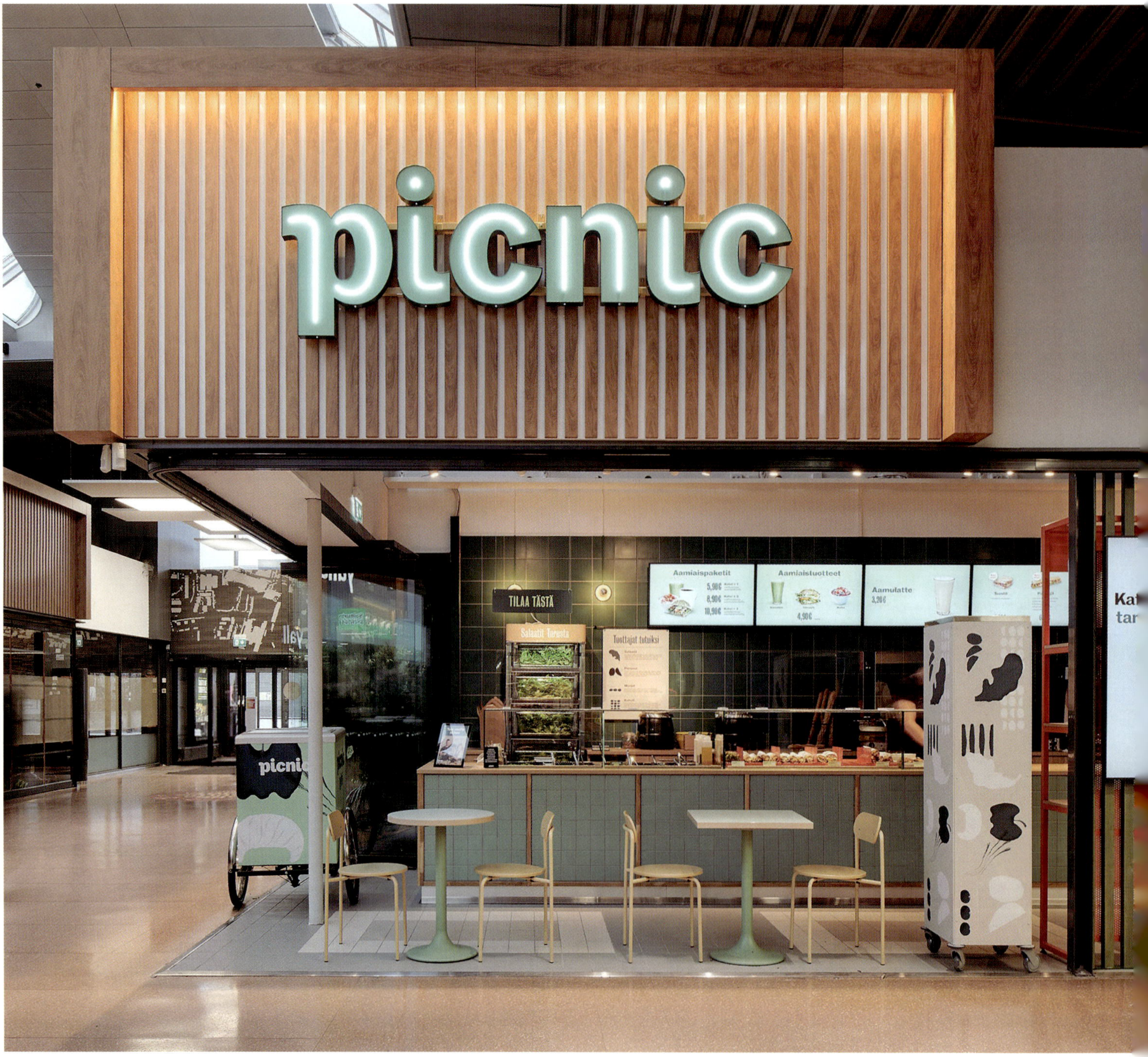
picnic
TILAA TÄSTÄ
Aamiaispaketit
Aamiaistuotteet
Aamulatte
picnic

pic nic
Meillä tehty Sinulle

pic nic

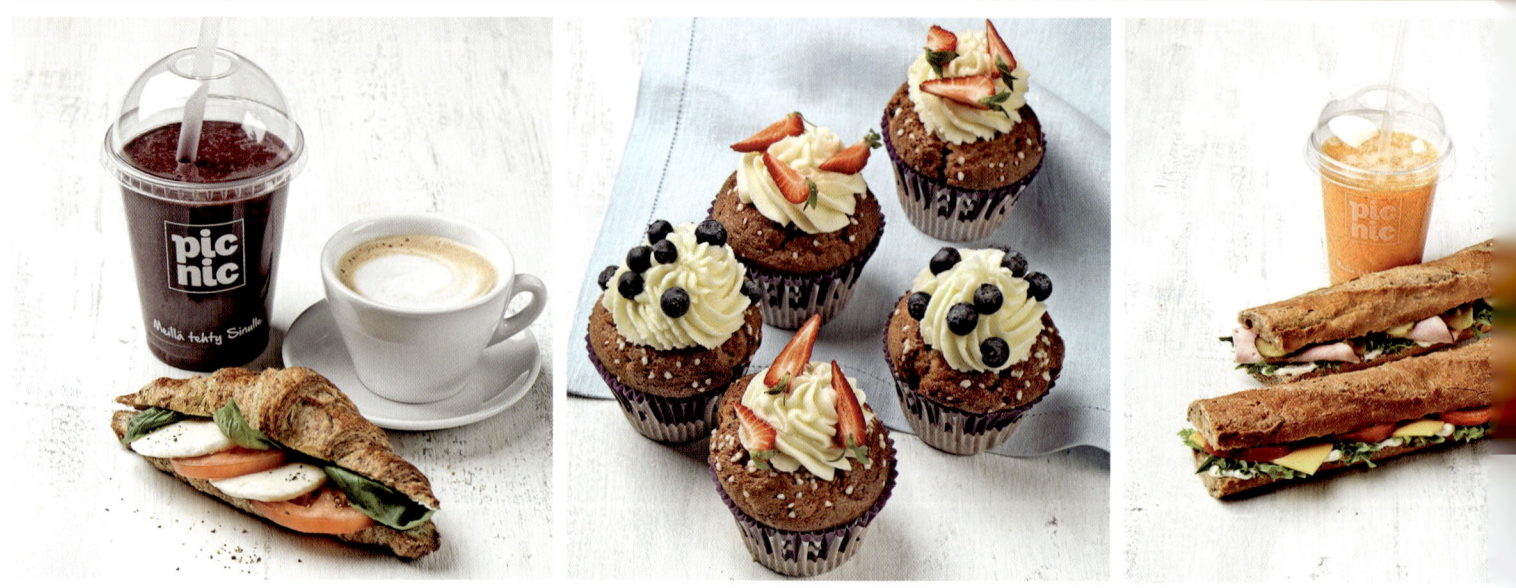

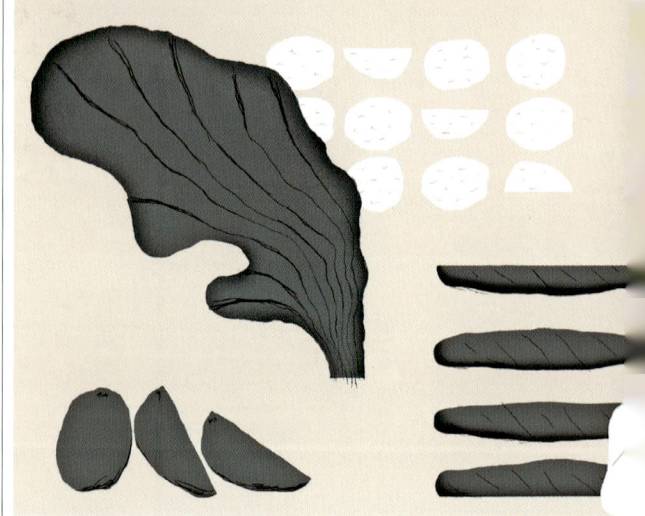

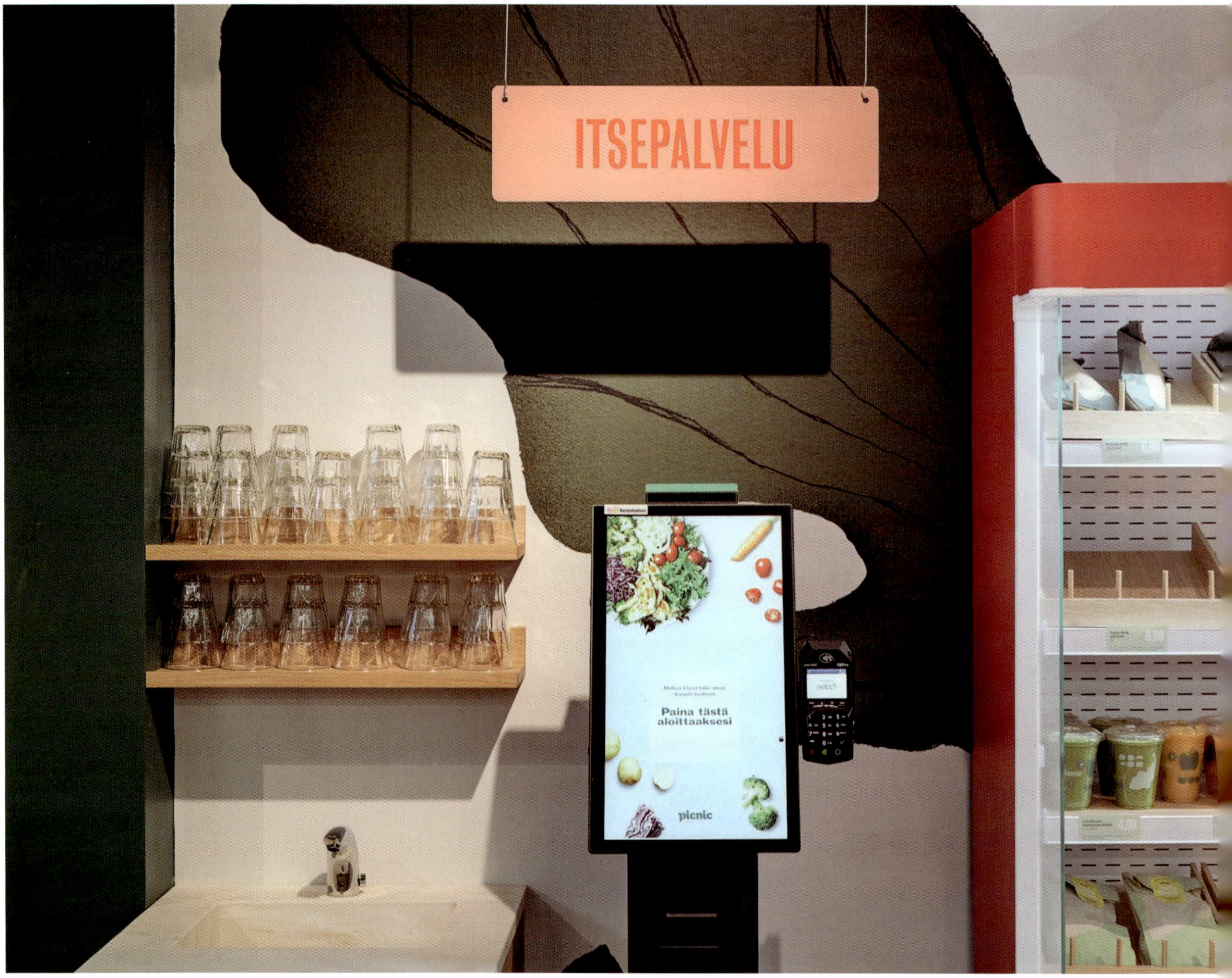
ITSEPALVELU
Paina tästä aloittaaksesi
picnic

ITSEPALVELU

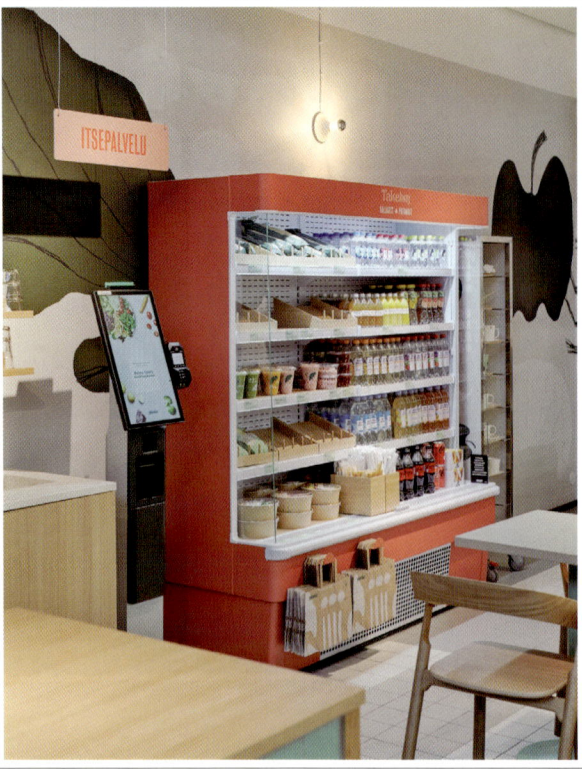

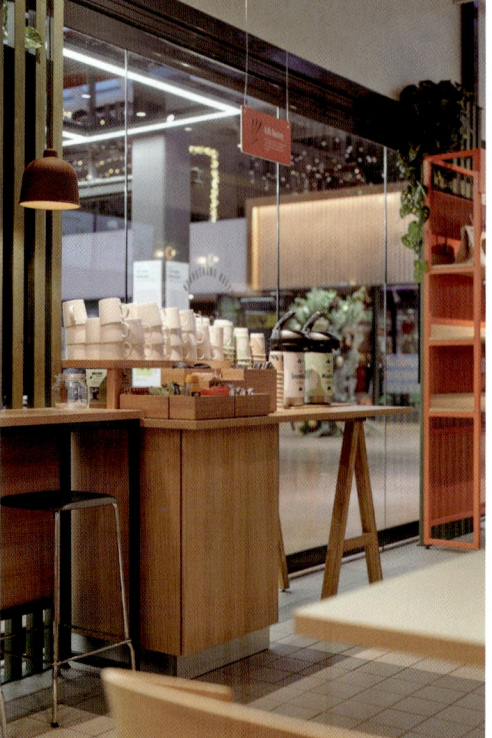

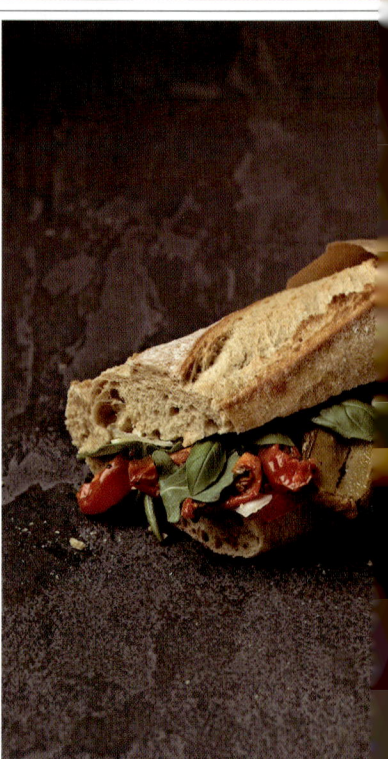

PICNIC: "Picnic's ambition is to be Finland's most passionate café company. Picnic companies consist of café-restaurant chain Picnic, founded in 1991 and known for its filled baguettes and baked potatoes, high-quality coffee focused La Torrefazione café chain as well as procurement and logistics company Europicnic. In total there are 49 cafés and the company's sales totaled €24 million in 2021. Sentica aims to develop and support the company in opening new cafés all around Finland. Eating outside of home is growing and interest towards the origin of food is driving the purchase decision making in increasing amounts. We believe that an excellent customer experience is built on responsible sourcing of ingredients, benefits of a wide chain and well-known brands.

Lingenhel Store
Vienna, Austria

Bakery, shop, bar, restaurant and cheese dairy - the "Lingenhel"

The "Lingenhel" project is unique in its own way and therefore not an everyday challenge for an interior design office like destilat. A visual and spatial identity was developed here for Johannes Lingenhel's culinary and gastronomic vision, for which there was originally no corresponding definition. The two cornerstones of this interior design concept are the listed building of the project on Landstrasser Hauptstrasse and Lingenhel's graphic corporate identity. Counters, bar counters and presentation furniture were inspired by the beams of the historic roof truss and are reminiscent of wooden beams stacked into cubes. Surface structure and haptics play important roles. Through the patination, they reflect changes that develop over time. This means that both the long history of the house and the manufacturing processes of certain foods find their place in the premises.

The central thought behind this interior design concept is not to fight against natural aging processes but to use them to increase quality. The original aging process thus becomes a refining process, which is often associated with tasty maturing cheese or raw ham. The bar becomes more and more beautiful over time through daily use - it is described with its own story. The second foundation of the architectural concept - the corporate identity developed by Germaine Cap de Ville for Lingenhel - is given a central role in the interior design with the Lingenhel check. The check is transformed from a purely abstract graphic pattern into the basis of the wall-mounted product presentation. Translated into the 3rd dimension, this develops into a shelf for the wine presentation.

The show dairy with adjoining tasting and event room in the old court stables is certainly a central area in the Lingenhel cosmos. The aim here was to combine the hygienic requirements of a cheese dairy with the requirement for an atmospheric, multifunctional event space.

Project details

▶**Design:** Destilat Design Studio
▶**Homepage:** https://destilat.at/en
▶**Area:** 286 m²
▶**Location:** Vienna, Austria
▶**Photography:** Monica Nguyen

About us

We spend most of our time indoors. Our spatial environment significantly determines our well-being. Our own home and workplace are particularly important. If you are still looking for the perfect interior that is personally tailored to you, we will be happy to support you. Your personal needs are just as important as our know-how.

A free initial consultation with us in our office serves to find out what possibilities may arise for you within the framework of interior design planning. This ranges from floor plan and furniture planning in the kitchen, bathroom and living area to detailed lighting planning and fine-tuning of colours and materials. And, of course, we will be happy to take care of the entire process for you.

The cheese dairy area was separated from the rest of the room by a glass wall and can only be entered through a hygiene lock. The sophisticated light dramaturgy is reminiscent of a theater stage and gives the cheese-making process the desired central role. The huge central table made of raw wooden beams and the two minimalist wire mesh chandeliers give the tasting room an archaic touch, which matches the character of the room and at the same time provides an attractive contrast to the sober, industrial atmosphere of the cheese factory.

The patina of time is celebrated in the wood beams, butchers-block counters, marble accents and horses' drinking trough. And, although not all of the materials are old, they have the appearance of patina and they will age beautifully with constant use. The plain cast-iron accents, the cool light fixtures and the minimalist color palette give the shop its sense of timeless yet modern charm. Our favorite aspects of this project are the curved form that repeats in the cast-iron accents, lighting, seating as well as the arched entryways and windows, and the wood ceiling that demands attention and draws the eye up. The in-house cheese makers will produce, for example, buffalo and goat's milk specialties including brie, camembert and mozzarella. The Lingenhel boutique also carries other delicacies such as prosciutto ham, preserves, breads and olive oil. – Tuija Seipell.

"Pleasure begins with taking your time. This applies to our products from the best aging cellars and pantries in Europe, which thus achieve their ideal maturity, and to the careful selection and preparation of honest, high-quality ingredients, as well as to maintaining relationships with our producers. However, we prefer to spend our time sharing our enthusiasm for excellent, handcrafted products with you - whether we are looking for a special gift in our shop, making cheese together with Robert Paget in the Stadtkäserei or as part of a successful evening in the restaurant and at the bar".

- Johannes Lingenhel -

The central thought behind this interior design concept is not to fight against natural aging processes but to use them to increase quality. Counters, bar counters and presentation furniture were inspired by the beams of the historic roof truss and are reminiscent of wooden beams stacked into cubes. Surface structure and haptics play important roles.

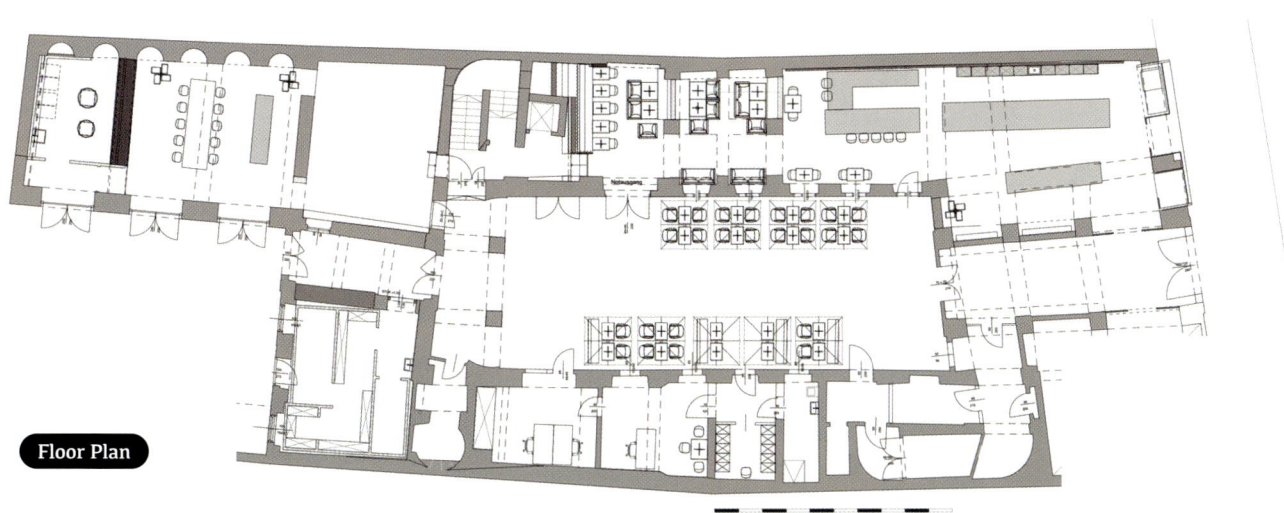

Floor Plan

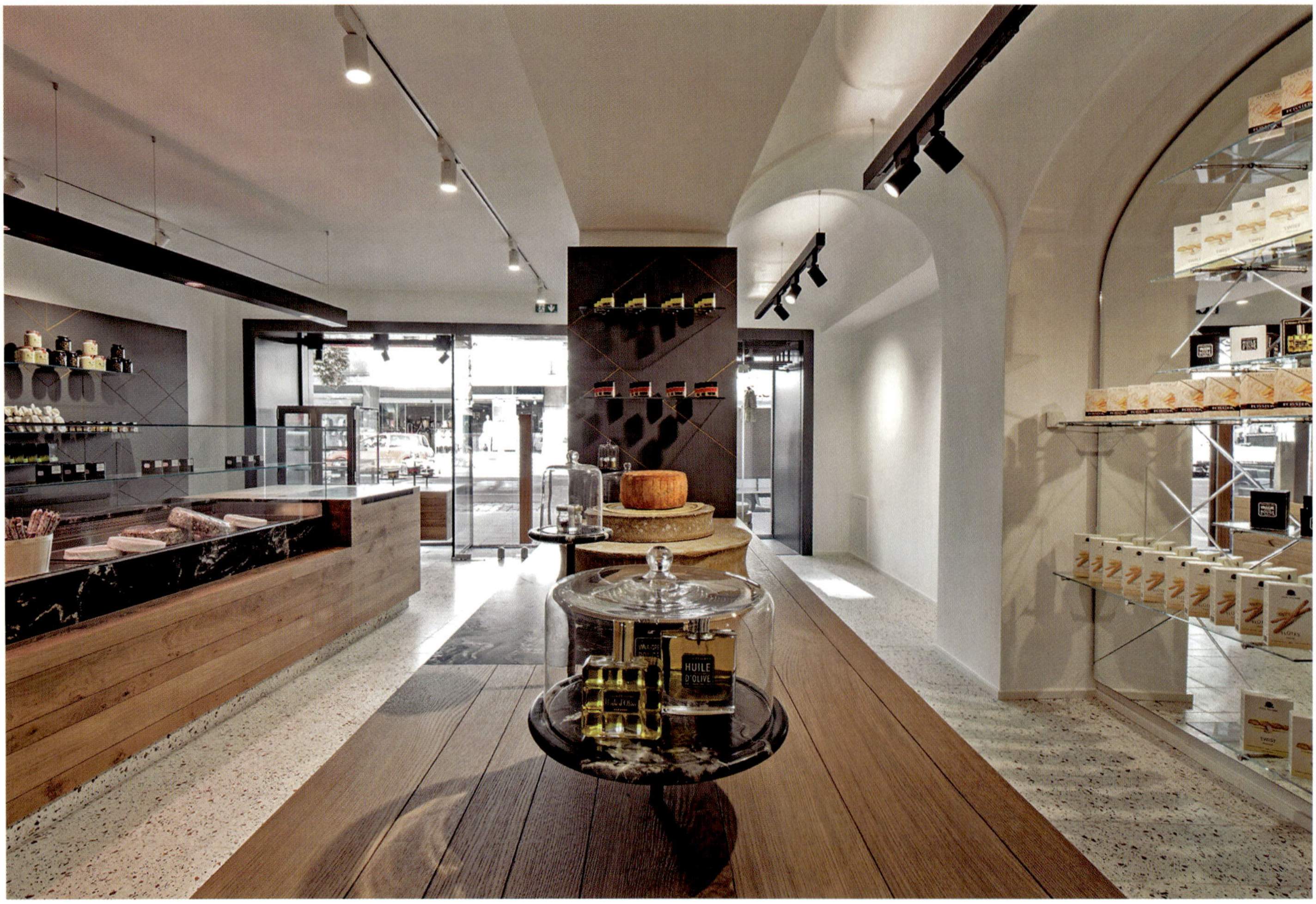
HUILE
D'OLIVE

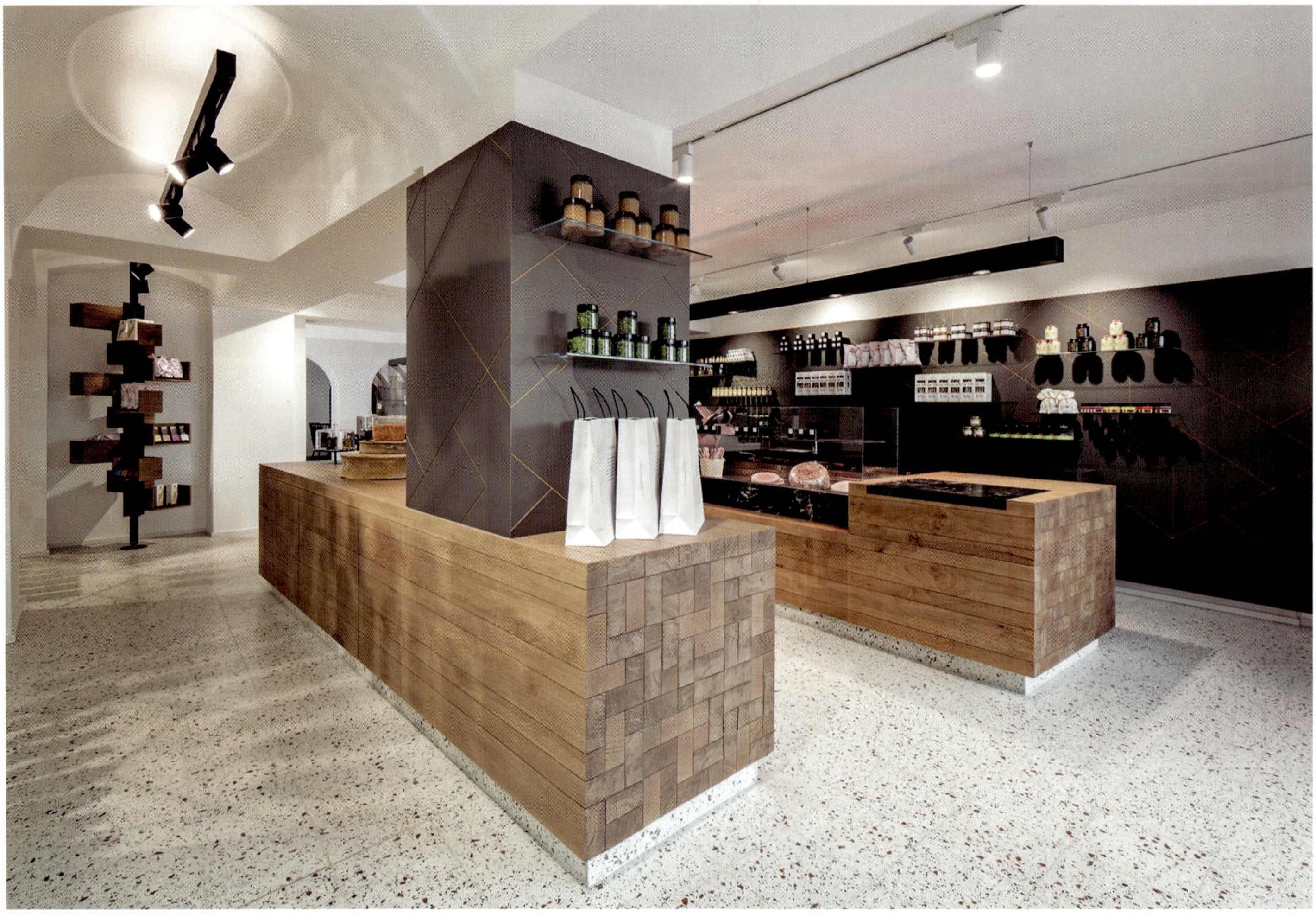

The check is transformed from a purely abstract graphic pattern into the basis of the wall-mounted product presentation. Translated into the 3rd dimension, this develops into a shelf for the wine presentation.

SHOP FIRST, UNWRAP GIFTS THEN-

Giving makes you happy. This starts with wonderful salumi and exquisite wines as thoughtful souvenirs, continues with sweet spices from the best manufacturers in Europe and does not end with the homemade creations from our gourmet workshops. In our shop you will find the right gift for even the most discerning connoisseur - of course also for yourself.

OUR CHEF UNDERSTANDS HIS CRAFTSMANSHIP -

Many think our chef has the best job in the world because he works every day anew with excellent seasonal ingredients fresh from the market, top products from the best manufacturers in Europe and our in-house cheese dairy. So it's only fair that he gives you the best lunch breaks, enjoyable evenings and unforgettable moments, we think.

-FRESH DAILY-

Our dishes tell the story of genuine food craftsmanship. It begins in the manufactories and maturing cellars of our top producers and takes its enjoyable course in our kitchen in combination with market-fresh seasonal ingredients. With so much enthusiasm for the product, it can happen that our guests leave knives and forks on the side.

Las Ramblas

Torun, Poland

Las Ramblas breakfast and coffee brewers & Bakery

▶Lingenhel Store: Mały Generała Władysława Sikorskiego 19, 87-100 Toruń, Poland / +48 511 464 51

BREAKFAST AND COFFEE BREWERS
Las Ramblas

Extended logo and branding for specialty coffee stores named "Las Ramblas breakfast and coffee brewers", distinguished by its constant focus on quality. We approached the project with a minimal attitude, designing a logo that refers to the sunrise, having the name of the brand in the central position of the sun and a super title spotted above, instead of the radials.

For the integration of the synthesis we used a balance of black and Pantone pink color 196C, both detected in to the overlord white. We paid special care to each and every application, so as to follow a common philosophy of simplicity and harmony, designing minimal forms and typography graphics. The official tagline of the brand is "love to coffee", inspired from their love to brew.

Project details

- ▶**Design:** Cursor Design Studio
- ▶**Homepage:** https://www.cursor.gr
- ▶**Area:** 30 m²
- ▶**Location:** Toruń, Poland
- ▶**Interior photos:** Michael Koronis
- ▶**Branding photos:** Zisis Dalakouras

About us

CURSOR DESIGN STUDIO® successfully accomplishes the communication of your business, services and products to the domestic and global market. We consider each new mission as a challenge for our professionalism, talent, enthusiasm and passion: recognizable characteristics through out the full range of our work. We face up branding as the ultimate communicative synesthesia and we faithfully serve all its functions: naming, logos, corporate identities, packaging, natural and digital promotion and art direction.
The company founded in 2002 by Apostolos D. Tsiovaras.

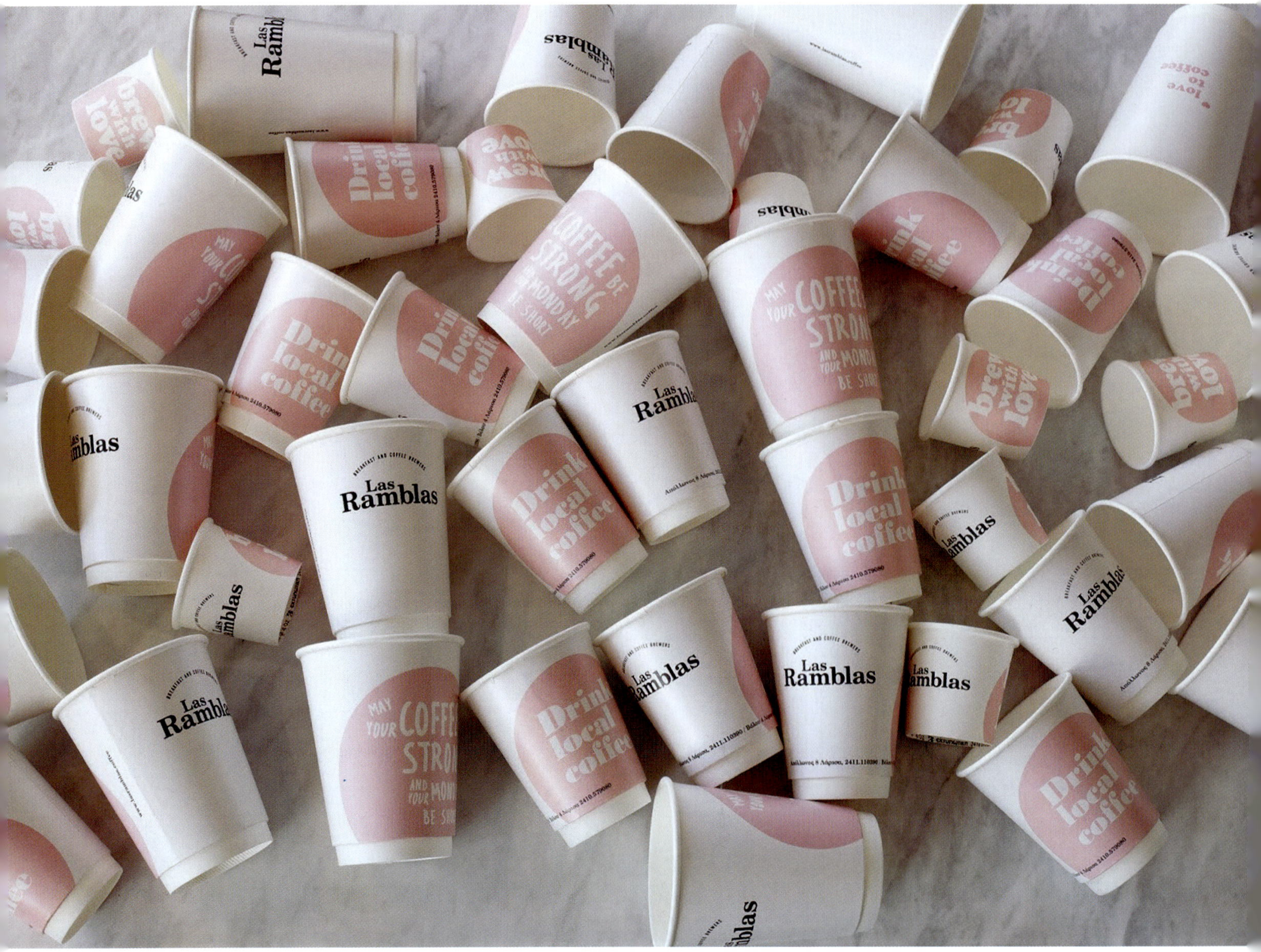
Las Ramblas
Drink local coffee
MAY YOUR COFFEE BE STRONG AND YOUR MONDAY BE SHORT
brew with love
love to coffee

Las Ramblas
Hungry...

BREAKFAST AND COFFEE BREWERS
Las Ramblas
PANAMA GEISHA

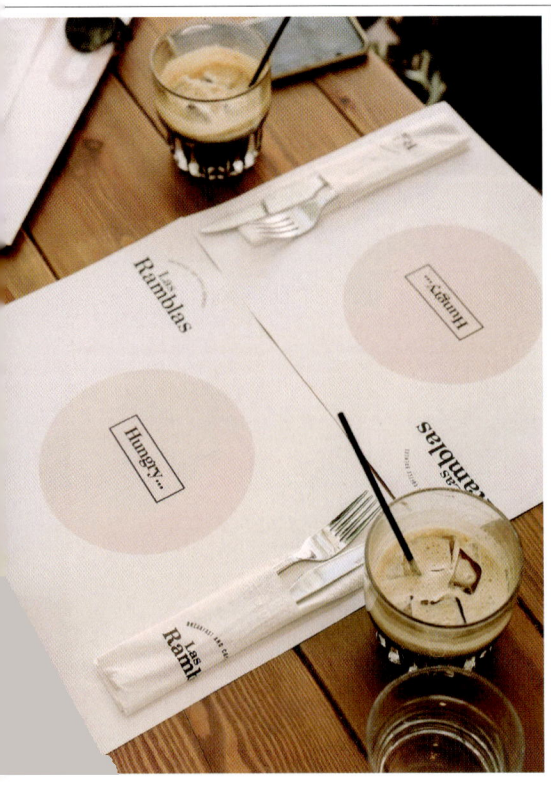

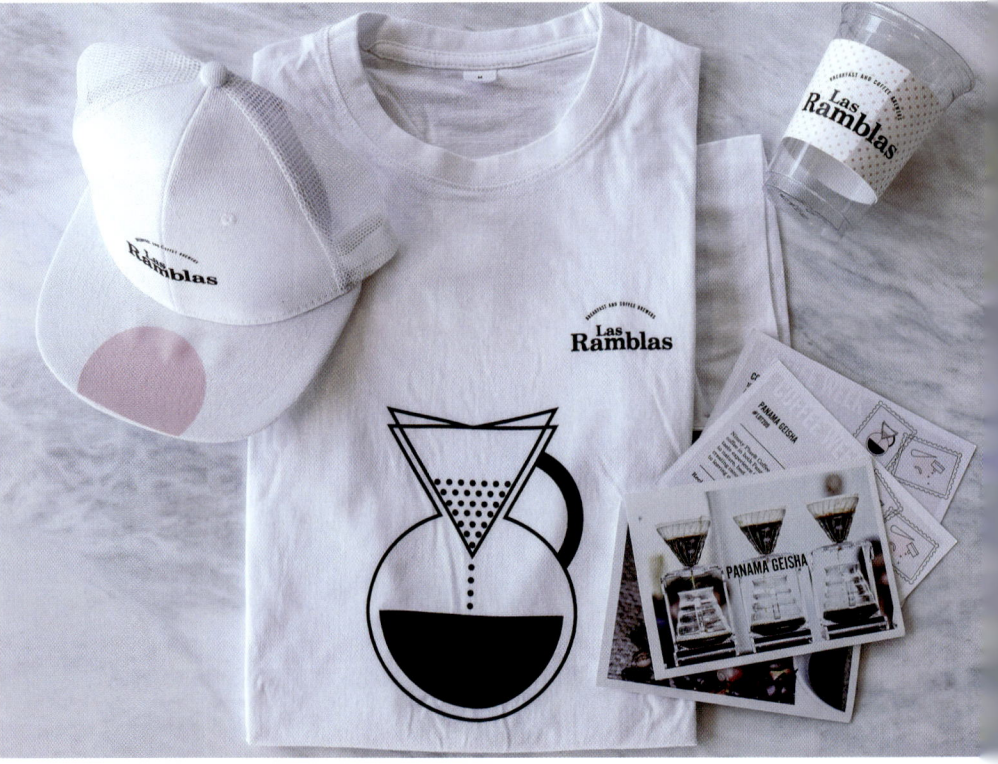

Toronto, Canada
Fuwa Fuwa

Japanese Pancake~Fuwa Fuwa - Golden Square

This modern café brings in youthful energy and fresh elements imbued with the steadfast qualities of traditional materiality. The spatial organization is a tribute to traditional alley streets of Kyoto, offering glimpses of what is yet to be explored under the playful yellow and white canopy.

Located in an outdoor plaza, the goal is to introduce the cafe's modern and forward presence to an otherwise traditional and subdued area. It is neither a grab-and-go cafe place nor a sit-down restaurant. It exists in the intersection of the two, and thus aims to combine the visual aesthetics of both; the brightness of the youth-led culture and the solid steadfast quality of a classic eatery. All the while, the design infuses visual traits of Japanese design to reflect the Japanese style dessert that it serves. Visitors are greeted by the soft curve and bright yellow of the high canopy as it frames a modern temple-like structure. Light emanates from it, softly brightening all corners of the space. It is flanked by corridors on either side. By entering the side valleys, two experiences become available; the hustle and bustle of pancake flipping and the cozy feeling of being tucked away in a small pocket of space. The narrow and long 168 square meter space needed to fulfill three key requirements: a space for delivery/pickup orders, privacy for in-store customers where they can converse and dine undisturbed, and an accessible and highly visible front of house.

Project details

- ▶**Design:** Studio Yimu
- ▶**Homepage:** https://www.studioyimu.com
- ▶**Area:** 168 m²
- ▶**Location:** Toronto, Canada
- ▶**Photography:** Scott Norsworthy

About us

Studio Yimu is a multi-disciplinary design firm based in Toronto, Canada dedicated in transforming out-of-the-box ideas into extraordinary modern spaces. Our expertise focuses on finding the perfect balance between ingenuity, art and technology to create efficient and functional designs.

Our team of professional designers will maximize the potential of your space, reflecting the desired atmosphere and message through unique elemental touches while ensuring that we honour the allocated budget. Become our client today and let us show you the different possibilities of you could do with your space!

The pandemic has made take-out and delivery more prevalent and it became essential to consider the distribution of space in relation to the flow of traffic. How can the travel of customers be minimized? How can this narrow space host distinct zones for different customers in an organic and passive fashion? The solution came in the form of highlighting the length of the space and creating a feeling of walking down a side alley street. By placing the sales counter in the front of the curved element, it effectively becomes the entrance, creating a simple dialogue of "inside & outside" between two contrasting zones.

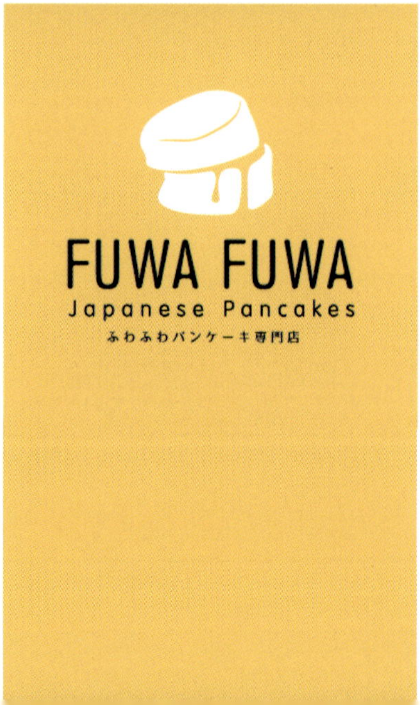

This café is one of the newest additions to a growing franchise with an existing brand identity. The challenge is maintaining their core identity while still creating a one-of-one space. We believe design is always built in context and therefore, should never be copy and pasted. We must look at the nuances of the space, recognize its unique abilities and limitations and tailor the design. The space is very long and narrow and the natural configurations options afforded to more rectangular spaces could be used here. This posed an interesting design challenge. The team drew inspiration from the history of the business.

The food techniques came to Toronto from the streets of Japan and we built the concept around the imagined idea of Fuwa Fuwa being a modern Japanese street valley juxtaposing the old and modern craft. We used simple design principles of emphasis, proportion, and placement to create focal point and distinct zones that still feel visually cohesive. Its beauty is not only in its visual appearance but also in the space's ability to provide comfort.

■ *By placing the sales counter in the front of the curved element, it effectively becomes the entrance, creating a simple dialogue of "inside & outside" between two contrasting zones.*

Gelato Dal Cuore

Shanghai, China

Cool balance of refined and classic which ties together the fun and playfulness of the ingredients.

Gelato Dal Cuore: 600 Shaanxi Bei Lu, Shanghai, China / +86 21 6148 1388

Gelato Dal Cuore is the latest addition to Jing'An eateries resurgence. When it comes to Gelato these guys really know what they are doing. using fresh imported ingredients to bring in traditional Gelato flavours with loads of seasonal and local twists. To compliment the high quality, Gelato Dal Cuore wanted some traditional elements that spoke to the heritage of classic Italian Gelato, mixed with modern elements and accents to complete their orange brand colours. The result is a cool balance of refined and classic which ties together the fun and playfulness of the ingredients. This shop is sure to have lines long into the Shanghai summer.

Project details

▶ **Design:** Hcreates design
▶ **Homepage:** https://www.hcreates.design
▶ **Area:** 90 m²
▶ **Location:** 600 Shaanxi Bei Lu, Shanghai, China
▶ **Photography:** Brian Chua

About us

Hcreates is an interior design and consulting studio based in Shanghai, designing projects across China and increasingly across Asia since 2010. The practice works with like-minded clients to achieve spaces that are designed to enhance one's environment and leave people feeling stimulated and inspired. Well-known in Shanghai for creating contemporary restaurants and bars, hcreates continues to develop a significant portfolio of office and health and wellness spaces too. Its design philosophy comes from an ingrained sense of kiwi ingenuity, innovation and practicality. Design should be simple, fun and clever.

▶ **Contacts** e-mail: info@hcreates.design / Call us: +86 21 6148 1388

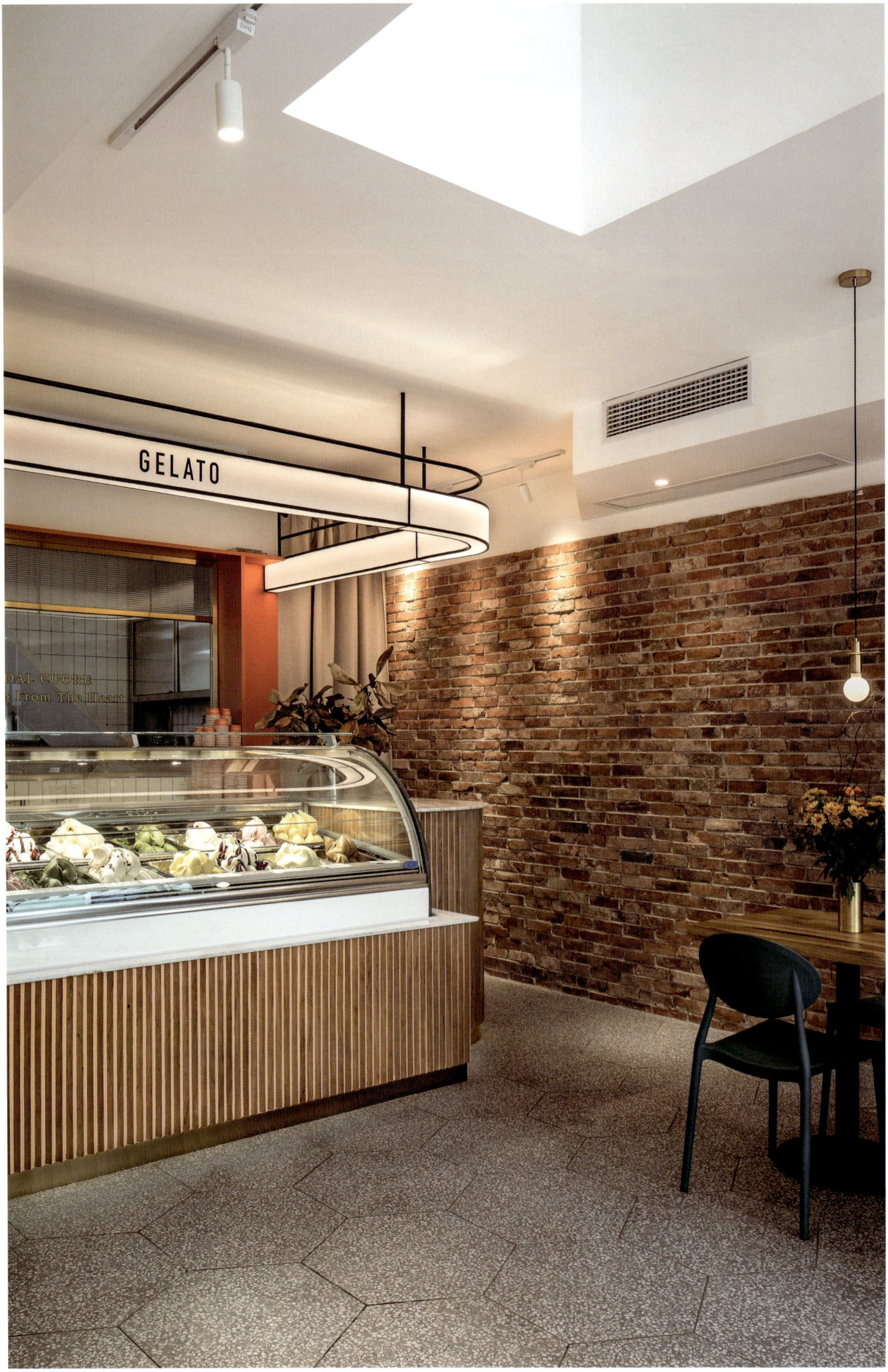
GELATO
DAL CUORE
From The Heart

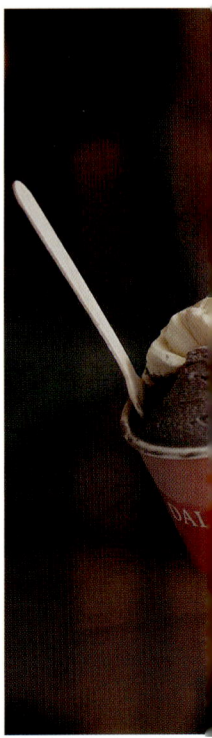

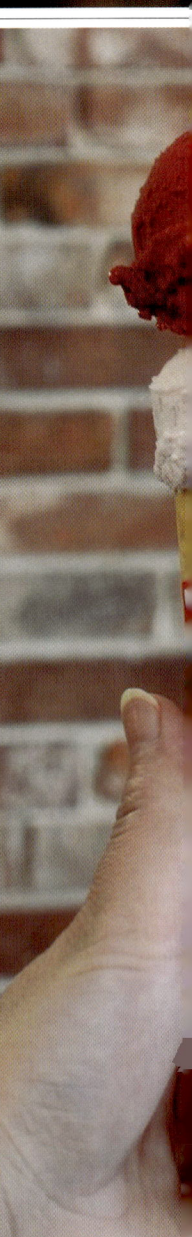

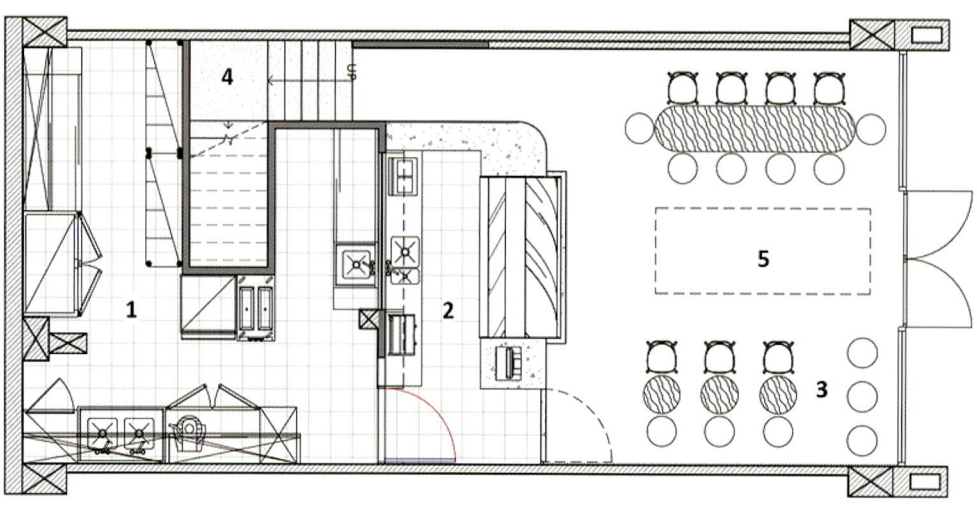

Fllor Plan

1. KITCHEN
2. GELATO COUNTER
3. SEATING
4. STAIRS TO OFFICE
5. VOID ABOVE

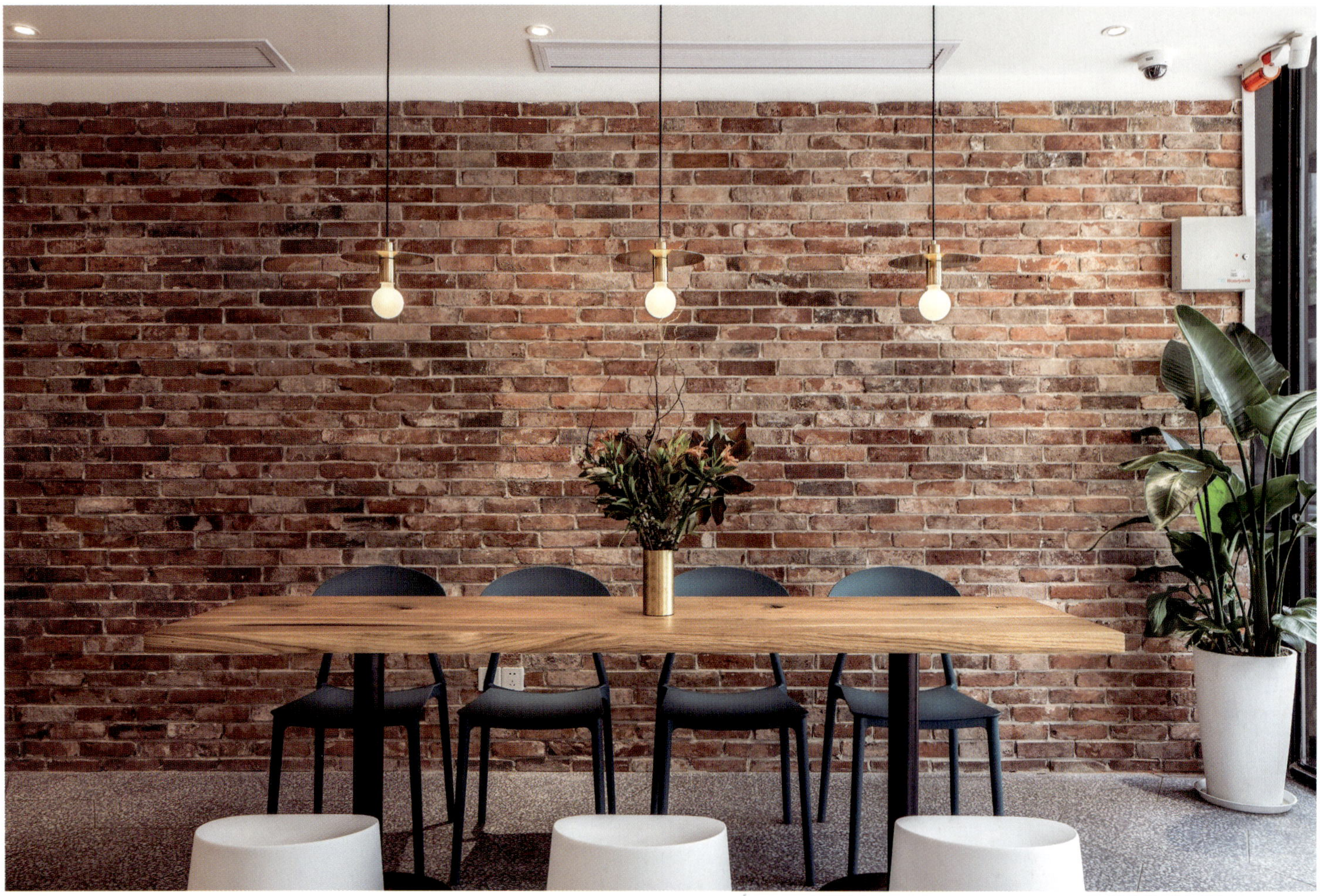

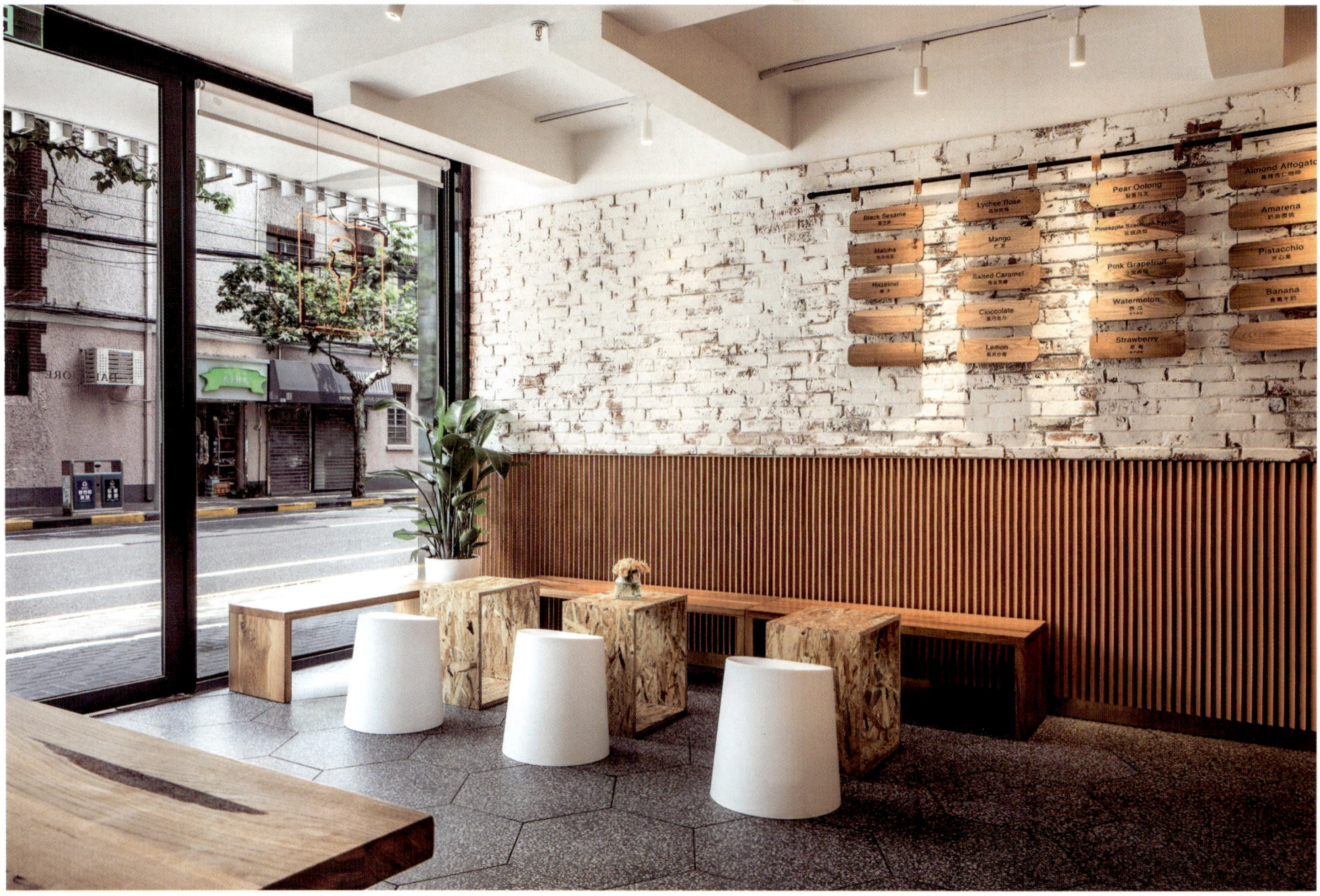

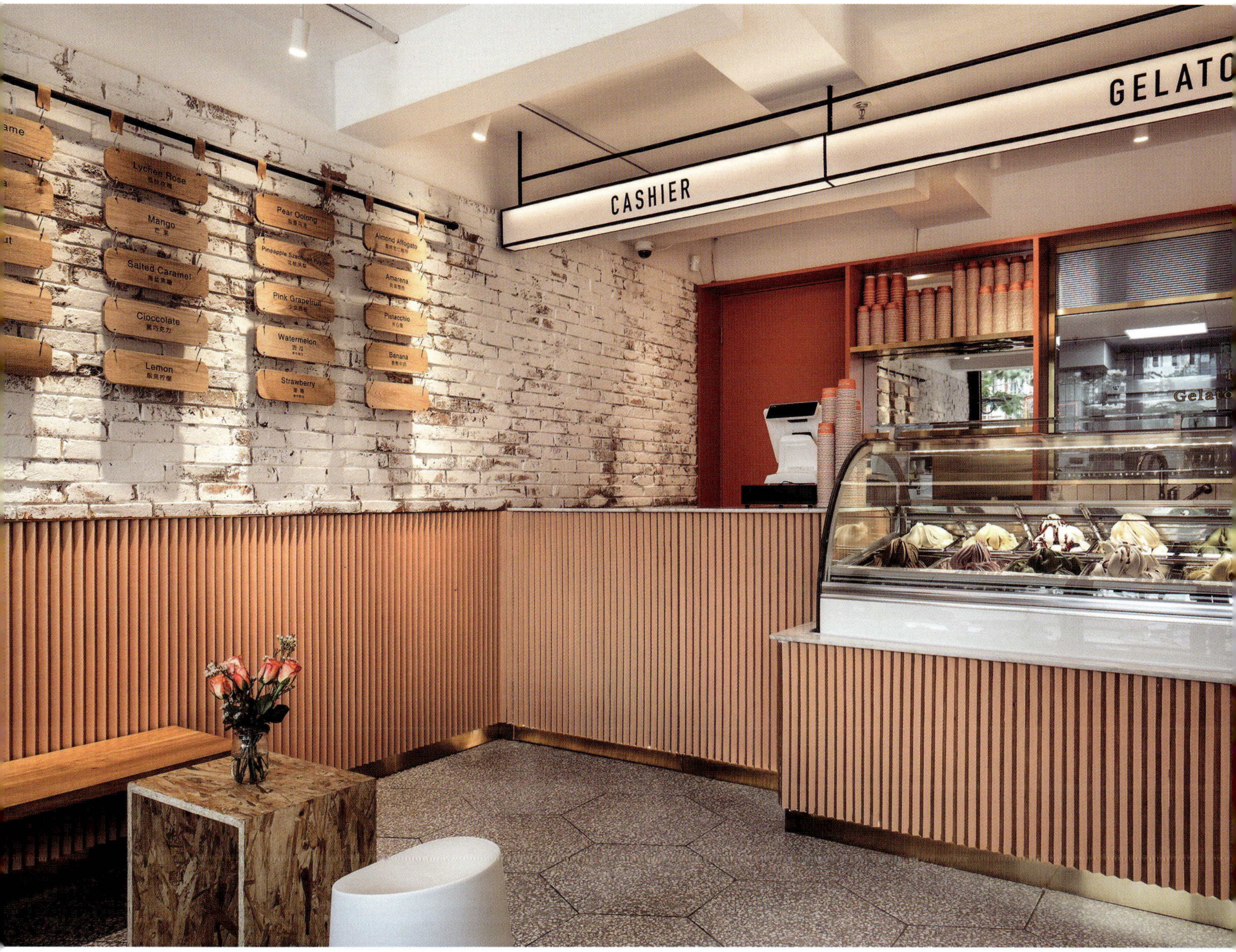
GELATO
CASHIER
Lychee Rose
Mango
Salted Caramel
Lemon
Pear Oolong
Pink Grapefruit
Watermelon
Strawberry
Pistacchio
Banana

gelato
DAL CUORE

gelato
DAL CUORE

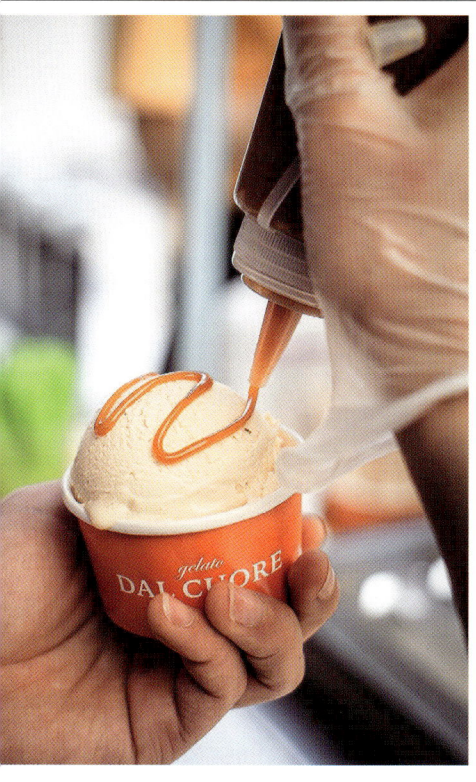

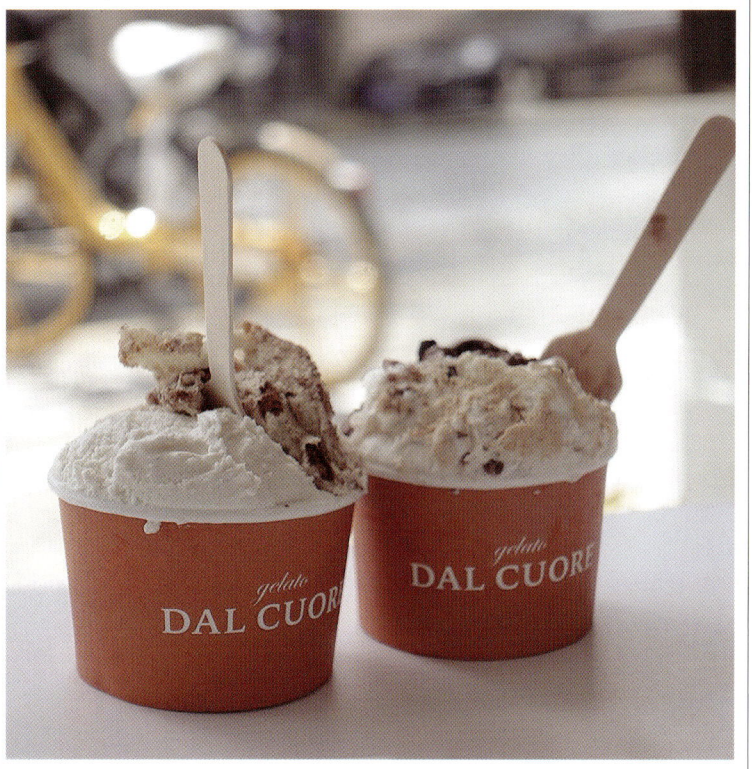

Gelato Dal Cuore: Serta, Portugal / +351 232 428 214

Serta, Portugal
Estrela Doce Bakery

The countless and exquisite details of a contemporary style decoration based on the plain tones and the unpretentious use of the materials.

Located over 20 years on the 25th April Street in the village of Sertã, Estrela Doce Bakery has just opened a new space on the same street but with a totally renewed image in a 125 m2 area. Considered a must-see reference in the Beira Interior region, the present owners took a step forward not only in product manufacturing but also in the space, always aiming to maintain its close relationship with customers inviting those responsible for their success to feel at home.

After conquering our senses, they invite us to stay between the countless and exquisite details of a contemporary style decoration based on the plain tones and the unpretentious use of the materials, like iron and birch wood, which convey into a unique harmony. The majestic windows let in the natural light that fuses with the light tone of the wood and the lamps, exclusively designed for this project, shine an indirect light on the tables, creating a warm and comfortable atmosphere. The use of marble from Estremoz gives elegance and sophistication to the space.

Project details

▶ **Design:** Sónia Triguinho
▶ **Homepage:** https://retaildesignblog.net
▶ **Area:** 156 m²
▶ **Location:** Serta, Portugal
▶ **Photography:** Mário Jacinto

About us

Sónia Triguinho, responsible for the interior design, imprinted comfort to the project, creating an intimate atmosphere with a contemporary expression. For the purpose, the designer privileged materials such as iron, wood, and marble, mostly with national production origin. Everything the eye catches is breathtaking. The showcase of cakes and bread makes us want to try everything. The smell coming from the manufacturing downstairs leaves no one indifferent.

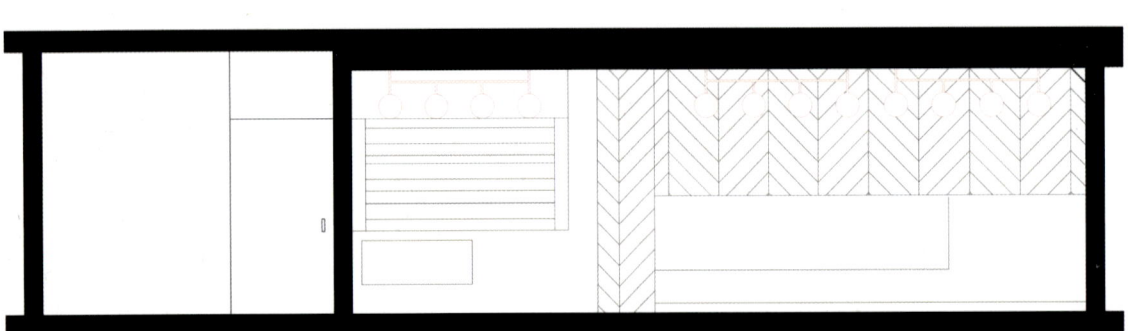

Section

Floor Plan

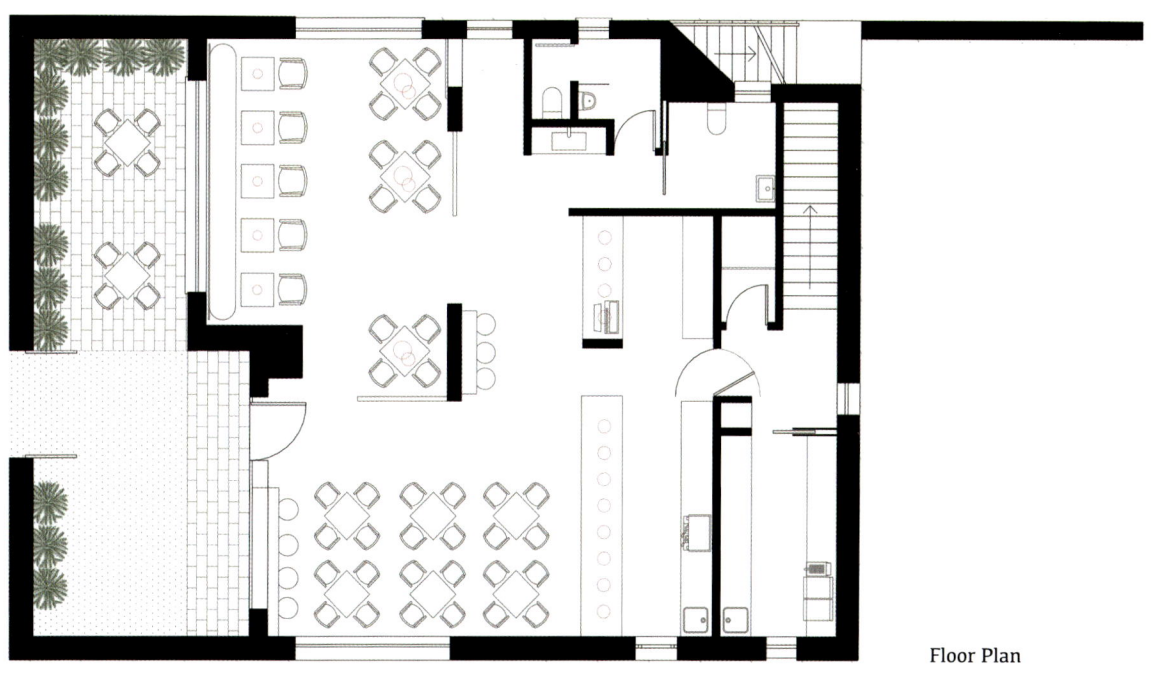

301